T0220235

USA Through the Lens of Mathematics

USA Through the Lens of Mathematics

Natali Hritonenko and Yuri Yatsenko

CRC Press
Taylor & Francis Group
Boca Raton London New York

CRC Press is an imprint of the
Taylor & Francis Group, an **informa** business
AN A K PETERS BOOK

Cover design by Olga Yatsenko

First edition published 2022
by CRC Press
6000 Broken Sound Parkway NW, Suite 300, Boca Raton, FL 33487-2742

and by CRC Press
2 Park Square, Milton Park, Abingdon, Oxon, OX14 4RN

ISBN: 978-1-032-13566-3 (hbk)
ISBN: 978-1-032-13349-2 (pbk)
ISBN: 978-1-003-22988-9 (ebk)

DOI: 10.1201/9781003229889

Typeset in Palatino
by MPS Limited, Dehradun

Contents

Preface.. vii
Preface for Students ix
Acknowledgments.. xi
Author Biographies xiii
Introduction xv

I The New Nation 1
I.1   The United States of America 1
I.2   Presidents of the United States of America............................ 13
I.3   The American Flag... 23
I.4   Symbols of the US States 31

II Geography of the United States.. 39
II.1   Geography of US States... 39
II.2   Highest Peaks and Lowest Elevations 47
II.3   Rivers and Lakes.. 56
II.4   Mysterious Places.. 63

III National and State Parks 71
III.1   National Parks.. 71
III.2   State Parks... 82
III.3   National Trails... 88

IV The US Highways 99
IV.1   Trips to National Parks 99
IV.2   Across the States... 107
IV.3   The US Highways... 119

V Constructions and Inventions in the United States 129
V.1   Architecture Wonders 129
V.2   Incredible Constructions... 137
V.3   Predictions and Reality.. 144
V.4   UNESCO Sites and High-Tech Companies............................ 150

VI The United States in Arts... 159
VI.1   The American Story in Art 159
VI.2   American Writers 166
VI.3   American Women in Arts.. 173
VI.4   Awards in Art 180

VII Shopping, Food, and Entertainment in the United States............**189**
    VII.1   Health Indicators..189
    VII.2   Food Production...193
    VII.3   Shopping in the USA...198
    VII.4   Walt Disney Parks..203

Appendix I **209**

Appendix II...**219**

Appendix III **221**

Index..**225**

Preface

Mathematics is the most beautiful and most powerful creation of the human spirit.

— **Stefan Banach,** *Polish mathematician*

This entertaining collection of 325 mathematical problems about American heritage was inspired by the students and educators the authors worked with.

The authors enjoy traveling around the United States and are fascinated with the remarkable US history, unique nature, and manmade constructions. While learning the astonishing stories behind these places and, through them, about the country they love, the authors always thought of their students and wanted to share this information with them (and others, too!) in an educational way that would also show off the magic and versatility of mathematics.

The authors hope that this creative approach will encourage deeper understanding of science and motivate simultaneous learning of both mathematics and social studies.

It is impossible to give comprehensive coverage of the whole story behind any country in one book; thus, the authors have not attempted to do so. Instead, they have tried to maintain the delicate balance between briefly summarizing an event and presenting a related mathematical word problem. The authors have worked diligently to ensure that their selected topics are both educational and enjoyable. Their main objective is to provide a thorough review of basic algebraic formulas taught in middle to high school. The presented collection of informative problems is carefully designed to make the book engaging and meaningful, but not too broad. Each individual problem has been created by authors with love, though some mathematical formulations may echo mathematical problems from the past.

The authors hope that the audience will enjoy reading this book as much as they enjoyed writing it. The authors welcome feedback and suggestions and will respond to all correspondence.

Preface for Students

Mathematics may be defined as the subject in which we never know what we are talking about, nor whether what we are saying is true.

— **Bertrand Russell,** *a British philosopher, mathematician, and Nobel laureate*

You probably enjoyed, and may be even agreed with the above quote by British mathematician and philosopher Bertrand Russell, didn't you? Perhaps you think that mathematics is a boring set of unrelated formulas irrelevant to life outside of class, right? We will not try to argue with you, but we challenge you to wait, take this fascinating book, and discover the interesting stories of US heritage and achievements integrated into 325 nontraditional mathematical problems. You will begin to understand the practical reasoning behind core mathematical concepts. Through solving these unconventional problems, you will uncover many remarkable facts. You will find out what might happen when people move with the "breakneck" speed of 15 mph or how to get a profit of 2,500 times the initial investment. You will be shocked to learn that both the telephone and television were once believed to have too many flaws to even be seriously considered; not to mention computers!

Learning about American presidents, flags, states, capitals, and other remarkable details, you will be better prepared for subsequent history, geography, science, and other classes. Solutions to the problems about National and State Parks, along with other unique US gems, may even help you effectively plan your future vacations.

In summary, this book was made for you and was written with respect and appreciation of your desire to learn and grow. Each problem has been thoroughly designed keeping you in mind and knowing your struggles with the subject matter. The authors have worked with students like you for many decades, are familiar with variations in mathematical backgrounds, and have felt the pain of your frustration. This book has been created to guide you through the challenges of algebra and, most importantly, to help you overcome fear of mathematics and reduce mathematical anxiety. Indeed, word problems are the heart and brain of mathematics. Solving them, you will gain mathematical knowledge and develop a deeper relationship with the science. You will quickly grasp the applications of and connections between algebraic formulas. You will be surprised by how fast your brain absorbs fundamental rules of algebra and geometry. None of the problems require any mathematical proficiency above algebra. They can be completed solely relying upon basic mathematical knowledge from middle and high school.

Remember three *do nots:*

- *Do not be afraid* if a problem looks too scary or difficult. Just relax and try to solve it. If you cannot solve it today, come back to this problem later. The magic will eventually work, and you will see that a complex formula can be reduced to a simple, but beautiful, answer. You will feel satisfaction and be proud of yourself – what a genius you are!

- *Do not give up* if a problem cannot be solved. Just keep trying or choose another problem! This book is full of problems to match any taste and any background. This book has the right problem for you, just look for it!

- *Do not skip a day* of solving at least one problem and learning something new. You will be enriched with new knowledge!

One final piece of advice to follow:

- Enjoy everything you are doing, and you will succeed. You can do this! You can be successful, and you will be! The authors *believe in you*!

Acknowledgments

No one who achieves success does so without acknowledging the help of others. The wise and confident acknowledge this help with gratitude.

– Alfred North Whitehead, *an English mathematician and philosopher*

The authors would like to express gratitude to their students and teachers for useful suggestions, inspirations, and ideas. A special thank goes to Mr. Philip Andrew, Ms. Morgan Roe, Mr. Quincy D. Lee, Ms. Crystal Holmes, Mr. Justin Tatum, Ms. La Nyrah White, Ms. Mikaela Dulan, Ms. Maria Alvarado, Ms. Krysten Maddox, Mr. DeAndre Reeves, Ms. Alondra Rosales, Mr. Michael Woods, Ms. Courtney Reyes, Ms. Alexandria Aguilar, and many other former and current students. Their help is incredible.

The authors appreciate help, advice, and friendly critique from their US colleagues – educators Dr. Fola Agusto, Dr. Benito Chen, Dr. Michael Nojeim, Ms. Anya Schmidt, Ms. Karim Seminario, Mr. Lucio Mera, Mr. Travis Teague, Dr. Jianzhong Su, Dr. Yunjiao Wang, Ms. Olga Yatsenko, Dr. Jeong-Mi Yoon, and international professors Dr. Raouf Boucekkine (France), Dr. Renan Goetz (Spain), Dr. Angels Xabadia (Spain), Dr. Nobuyki Kato (Japan), Dr. Seilkhan Boranbayev (Kazakhstan), Dr. Sergey Lyashko (Ukraine), Dr. Ludmila Nikitenkova (Belarus), and many others. The authors are also thankful for support they receive from President Ruth Simmons, Dean Dorie Gilbert, Dr. Supranamaya Ranjan, Dr. Victoria Hritonenko, Dr. Maria Leite, Dr. Inna Nosko, Ms. Glenna Sharon, and other friends and colleagues.

The authors would like to especially acknowledge constant assistance and helpful comments from the book editor Mr. Robert Ross.

While researching the milestone and pivotal moments presented in this book, the authors used a wide array of wonderful electronic reference tools, such as Google Search, Wikipedia, National Park webpage, and other Web sites. The authors are sincerely grateful to professionals for creating and maintaining these web sources and making them available to the general audience.

Author Biographies

Dr. Natali Hritonenko is an award-winning professor of mathematics and Associate Dean at Prairie View A&M University. She has shared her research results with a diverse team of experts at the leading centers and universities in the USA, Belgium, France, Spain, Japan, Russia, and other countries. Dr. Hritonenko has authored 7 books and over 150 papers on her research in mathematical modeling and education. Some of her books are used as textbooks around the world and translated to other languages. She is also on the editorial board of 9 international interdisciplinary ranked journals. She is continuously working on improving mathematical education. Dr. Hritonenko is the founder and director of PVAMU REU Site: Mathematical Modeling in the Sciences supported by NSF.

Dr. Yuri Yatsenko holds a PhD from Kiev State University (Ukraine) and a Doctor Hab. Degree from the USSR Academy of Sciences (Moscow). He was a professor at science, technology, and business schools in Russia, Ukraine, Poland, Canada, and the USA. For five years, he held senior positions in data analytics and operations research at international companies in the USA and Canada. Since 2002, he has been a professor of quantitative methods at Houston Baptist University. He has published 8 books and more than 200 research papers in top-ranked scientific journals.

Introduction

Mathematics is the language with which God has written the universe.

— **Galileo Galilei,** *an Italian astronomer, physicist, and engineer*

Mathematics courses are among the most challenging subjects in high schools, colleges, and universities. They are considered a barrier for successful studies by most students around the world. Sadly, insufficient knowledge of basic mathematical concepts, which essentially are the alphabet of science, is hindering students in their desired academic and career paths. Students continue to struggle in mathematics, physics, chemistry, biology, engineering, business, and other classes. The fear that many students experience when seeing a mathematical problem, especially a word problem, blocks their mind and prevents them from thinking and, even, reading the problem thoroughly. Such students do not believe in their success and get stuck, because they know in advance that they cannot do it.

Various teaching strategies have been developed to enhance students' mathematical background. One of novel educational strategies to lessen struggle with mathematics is to create surprising problems that can spark interest and desire to solve them. Integrating such problems into a multi-disciplinary cross-cultural content enhances a mathematical curiosity and boost development of strong mathematical and social skills. Indeed, interdisciplinary research is a major trend in modern educational curriculum. This entertaining book aims to assist this goal by offering thoroughly designed mathematical problems inscribed in applied content. Readers will review algebra fundamentals while learning hundreds of fascinating facts about the United States.

This book has been inspired by the authors' students, schoolteachers they work with, and participants of their numerous special sessions and workshops on innovative educational techniques. The authors have received many requests to share such problems and explain how to design them. They hope that this book will provide encouraging examples of how to merge mathematical problems with phenomena or events.

Unique Features of the Book

For the things of this world cannot be made known without a knowledge of mathematics.

– **Roger Bacon,** *a medieval English philosopher and scholar*

This enjoyable book demonstrates an innovative educational approach to simultaneous multidisciplinary and mathematical learning. It is *unique* among mathematical books in its devotion to reflect stories from the remarkable US heritage through the lens of mathematics. This book is neither a basic or comprehensive textbook with examples, nor a book with mathematical problems related to certain topics. It is, rather, an effort to combine both absorbing new information and improving mathematical culture. One of the key features of the book is its unexpected use of mathematics. The element of surprise will trigger a positive emotional response which, in turn, will make the reader intrigued and inspire curiosity and reflection.

The authors attempt to cover a considerable amount of algebraic knowledge and mathematical statements in one book. Its problems can be effectively used in a classroom or completed during leisure time.

The multidisciplinary character is another highlight of this book that distinguishes it from other mathematical books. Indeed, nature does not recognize the borders between disciplines artificially created by humans. New discoveries are often made on fringes of currently defined scientific areas. Cross-disciplinary vision has been the priority of successful research outcomes since ancient times. Solving captivating mathematical problems will lead to answering *Where?*, *When?*, and *Why?* questions on the history, geography, and achievements of our great nation. This interdisciplinary book contributes to innovative teaching strategies that not only improve students' analytical and problem-solving skills, but also broaden their scientific vision and cross-cultural awareness.

Audience

> We will always have STEM with us. Some things will drop out of the public eye and go away, but there will always be science, engineering, and technology. And there will always, always be mathematics.

> — **Katherine Johnson,** *an African American mathematician*

The authors aim to highlight the applicability of mathematics and emphasize multifaceted approach in education. The book can be used as a great supplementary text for universities, colleges, and schools in different classes, from chemistry to history and literature, and, of course, mathematics. Knowledge of basic algebraic formulas is the only mathematical prerequisite for solving suggested problems.

The book offers middle and high school, college, and university students a rich collection of attractive mathematical problems that will assist them in developing STEM skills, enriching their mathematical literacy, and, simultaneously, boosting their knowledge and appreciation of the US heritage. Practicing with these problems, high school seniors will become better prepared for their ACT and SAT exams and various future college courses. Indeed, expert reports demonstrate that students with appropriate mathematical preparation from high school show better performance in any college major they pursue.

The variety of suggested problems make them suitable for school or college mathematics clubs, different science competitions and Mathematical Olympiads.

Working on these applied mathematical problems, future mathematics teachers and college seniors will review basic algebraic concepts and master their mathematical skills, which will equip them with basis needed for successful passing their teacher certification, GMAT, MCAT, GRE, and other exams to enter a graduate school.

School curricula are shifting toward multidisciplinary education in a rapidly changing world. This book will be undoubtedly valuable for students and teachers in the development of cross-cultural and cross-disciplinary communications and motivate the global awareness as well. The presented problems are a great help to mathematics teachers and instructors who are looking to enliven their class with readymade entertaining problems or are working on their own collection of nontrivial problems related to various topics. The teachers would benefit from familiarizing themselves with novel interdisciplinary education trends and bringing these ideas to their classrooms. Problems can be selected or edited to fit any students' background and mathematical topic.

The book is ideal for college instructors and schoolteachers who develop interdisciplinary studies and seek innovative ways to demonstrate the relevance of mathematics to other subjects. They can find a suitable problem to be added to most courses in the standard high-school and college undergraduate curriculum. Detailed solutions and different ways of solving a problem included in the book will be a great asset for beginning mathematics teachers and instructors that teach other disciplines.

Finally, the authors hope that this book will be interesting for everyone who loves adventure, challenge, and enjoys mathematics.

Structure of the Book

Mathematics is the music of reason.

— **James Joseph Sylvester,** *an English mathematician*

This entertaining book consists of seven chapters, three appendices, and an index.

Its seven *chapters* are further divided in three to five themed sections. Each *section* starts with a brief introduction for its mathematical problems. In addition, a brief outline of the facts or events to be discussed precedes each problem or a set of problems. An applied story for a section or a set of problems is distinguished by appearing in italicized formatting. A mathematical problem and its plot are separated by a blank line space. All problems can be solved separately and independently of their applied descriptions. However, the mathematical problems complement their related story and are recommended to be solved to reveal remarkable details presented there. While all problems in one section are not always directly connected, they fit its main concept. The great variety of problems aims to both reveal incredible legacy of the United States of America and furnish basic algebraic statements at the same time.

The *answer key*, followed by *detailed solutions*, conclude each section. Several ways of solving a problem are presented in cases when each way involves either fundamental techniques or nonstandard methods that significantly simplify the solution. Numbering of formulas is avoided to make solutions more appealing to readers.

Appendix I offers a concise list of basic formulas and theorems needed to solve problems from this book. References to statements presented in Appendix I are provided in a *solution key* and can be easily found. For convenience, its sections start with the first letters of the property or rule. For instance, Appendix I-QE refers to solving or analyzing quadratic equations.

Appendix II links mathematical topics to book problems. *Appendix* III matches the US states with problems where these states are mentioned.

In summary, if you:

- Forget a mathematical formula or statement, then go to Appendix I
- Look for a problem with a certain mathematical content, then go to Appendix II
- Need more information on the US state, then use Appendix III

The *Index* is a useful guide to all mathematical concepts, relevant facts, and other entries in the book. They are listed in alphabetical order.

Mathematical Problems

To solve math problems, you need to know the basic mathematics before you can start applying it.

— **Catherine Asaro**, *an American science fiction author*

The collection of 325 problems is designed to fit any ability, background, and taste. A brief, italicized introduction accompanies each problem or set of problems. Following the recommendations of the first readers, numbering of formulas is avoided to make explanations clearer and friendlier for a general audience. Proposed problems are of different styles, types, and complexities. Some problems may look very similar, though a small variation in their formulation changes the level of difficulty and may require a completely different solution strategy. Moreover, some problems cannot be solved based on the information provided.

There are five designs of problems:

- *Puzzles* to be solved to reveal some information.
- *Word problems* with given equations to solve or mathematical statements to simplify.
- *Word problems* that require designing and solving a mathematical equation, inequality, or their system.
- *Multiple-choice problems* that offer several possible answers to choose from.
- *Problems with hints* that provide additional information as extra clues-tips. The solution (not answer!) can vary depending on the selected hints. It is important to explain why, where, and what hints should be used. A hint can
  - Lead to a statement needed to solve a problem
  - Be reduced to statements already presented in the problem
  - Help find a unique solution because other hints lead to either no or many solutions
  - Be useless and not applicable to the problem

Independently of their design, the problems can be of varying levels of complexity. Following suggestions from the first readers of this book, the authors have removed notations of complexity levels to avoid unnecessary fear of incapability of solving, or, even, starting, a difficult problem.

There are five levels of complexity:

- *Simple problems* do not require much mathematical training. They even can be figured out without any mathematical formulas, though their algebraic solution is provided as an alternative method of solving the problem.
- *Easy problems* involve the straightforward application of algebraic operations, formulas, or identities. These problems avoid complicated calculations and involve little mathematics.
- *Moderately difficult problems* include manipulating several mathematical formulas, performing several steps, or simplifying a complex formula to a simple and beautiful answer that reveals some interesting fact.
- *Nontrivial problems* involve not only the straightforward application of algebraic formulas and their combinations, but also logical thinking,

e.g., choosing an appropriate answer if not all of the necessary information is provided, or several solutions are possible.

- *Challenging problems* expect deeper thinking, critical reasoning, and thorough investigation to reveal interesting mathematical connections and complete the solution. In some cases, it is necessary to choose an option with relevant, practical meaning. A certain level of mathematical maturity is desired to solve the problems of this group.

The authors hope that this collection of problems can be effectively used to enhance mathematical knowledge and skills.

An answer key and detailed solution accompany each problem. They are presented at the end of each section. To illustrate the variety of mathematical techniques, different ways of solving a problem are often shown and thoroughly analyzed. Reasoning of hint selection is justified and discussed for problems with hints.

Finally, the authors studied various sources, reference works, and websites to collect information for the book. In some cases, the sources report slightly different data. Despite this inconsistency, the problems provide a reliable general description of real stories and events. As English American syndicated cartoonist Ashleigh Ellwood Brilliant jokes, *My sources are unreliable, but their information is fascinating.*

I

The New Nation

I.1 The United States of America

A new land was discovered or, probably, rediscovered by Christopher Columbus in 1492 and appeared on a world map in 1507 as the lands of the Western Hemisphere America. The land was named after the Italian explorer Amerigo Vespucci. Thirteen British colonies declared their independence from Britain in 1776, and the United States of America was born.

1. The United States of America is the union of 50 states and a federal district. It has populated and unpopulated territories in the Pacific and Atlantic.

   The difference between the fourth powers of the numbers of unpopulated and populated US territories is 5936, while the difference between their squares is 56. How many unpopulated and populated territories does the US have?

2. The World War II Memorial in Washington is an important symbol of American national unity. It was open to the public on April 29, 2004 and joined the National Park System later that year. It honors 16 million Americans who served during World War II. Each of 4048 Gold Stars on the Freedom Wall represents 100 Americans who died in the war. Its two triumphal arches highlight the US victory on the Atlantic and Pacific fronts. Each pillar of granite columns symbolizes the nation's unity and is engraved with the name of a state, federal territory, or the District of Columbia as of 1945.

   How many pillars are in the World War II Memorial? How many states and federal territories did the US have in 1945? Name these federal territories. Use the following hints to answer the questions.

   A. The number of the US states in 1945 is the same as the number that represents the area of an isosceles triangle $\triangle ABC$, $AB = BC$.

DOI: 10.1201/9781003229889-1

B. The line segment AD in $\triangle ABC$ connects the vertex A with a point D on BC and divides the area of $\triangle ABC$ in halves.

C. $AB + BD = 15$ units and $AC + CD = 17$ units.

D. The number of the US federal territories in 1945 is one less than the value of the height from the vertex B to the side AC.

3. Delaware became the first state on December 7, while Hawaii is the most recent US state. The years when Delaware and Hawaii became US states have a lot in common:

   A. Neither is a leap year

   B. The second and last digits in each year are the same

   C. The sum of all digits of the year when Delaware became a state is 23

   D. The sum of all digits of the year when Hawaii became a state is 24

   E. Only the first digit 1 appears in the years of both states

   What years did Delaware and Hawaii join the USA?

4. Pennsylvania became the n-th state in 1787, where n is the number of all factors of the year. Illinois became the m-th state in 1818, where m is the number of all factors of the year if the number of all factors is written in the reverse order. What are the statehood orders of Pennsylvania and Illinois?

5. When the Civil War began, the western region of Virginia declared its independence from Virginia, joined the side of the Union States, and became a state 2 years later, on June 20.

   The last (units) digit of the year when West Virginia was admitted to the union is the two-digit number formed from the first two digits of the year divided by the third digit. The sum of the first two digits of the year is greater than 7. The difference between the current rank in size of West Virginia among other US states and the statehood number of West Virginia is the third digit of the year, while their sum is the product of the last digit of the year and the two-digit number composed of the first digits of the year after both are increased by 1. What year did West Virginia become a state? What is the statehood number of West Virginia? What rank in size does West Virginia have?

6. The last three digits of the year when North Carolina joined the nation and became one of the original 13 states form an arithmetic sequence. Different values of x that satisfy the equation

$$xy - 3y + 6x - x^2 = 29$$

produce the second and the third digits of the year, while the value of y gives the last digit. Find the year when North Carolina became a state.

7. The last three digits of the year when North Carolina became one of the original 13 states form an increasing arithmetic sequence with the common difference of 1. The last two digits of the year satisfy the equation

$$xy - 3y + 6x - x^2 = 29,$$

where x stands for tens and y for units. Find the year when North Carolina became a state.

8. The United States Congress established the Missouri Territory in $1\overline{a}cb$. It was the westernmost border of the country. Missouri became a state in $1\overline{a}bc$. Find the years when the Missouri Territory was established and became a state if a, b, c, are coefficients of the equation of a circle

$$x^2 + y^2 + ax - by + c = 0$$

with the center at $(-4,1)$ and radius of 4.

Remark. $1\overline{a}cb$ represents a four-digit number, e.g., 1854.

9. How many states joined the Union in the year that is 151 times the number composed of its last two digits, which in turn is the sum of all digits of the year? The number of the states is the third digit of the year. What is the year? Can you name these states?

10. How many states joined the Union in the year that is k times bigger than the number formed by its last two digits, which is also the sum of all digits of the year? The number of states is the third digit of the corresponding year. The first digit of the year is 1. The year is divisible by 6.

    Can the problem be solved for any k? Define all values for k for which the problem has a solution, a unique solution? Name these states.

11. Either none, one, or several states joined the Union in this year. The number formed by the first two digits of the year divided by the last digit is the sum of the first digit and the absolute value of the difference between the second and the last (units) digit of the year. The first digit of the year is 1.

    Find the years when states joined the Union if the third digit of each year satisfies one and only one of the following hints:

1. The third digit is either a perfect number or a number that can be divided by any number but cannot divide any number

2. The third digit, which is odd, is among Lucas numbers or perfect squares but not both

3. The third digit is a Fibonacci number not divisible by 3

Remark: The solution is not unique. There are several possible options.

12. The sum of the first two digits of the year when Texas became a state is equal to the sum of its last two digits. The two-digit number obtained from the last two digits is related to the two-digit number obtained from the first two digits of the year as 5 to 2. What year did Texas become a state?

13. The sum of the first two digits of the year that Texas became a state is equal to the sum of its last two digits. The two-digit number obtained from the last two digits is related to the two-digit number obtained from the first two digits of the year as 5 to 2. The first digit of the year is 1. When did Texas become a state?

Use the following hints if necessary or explain why they do not bring any additional information.

1. The year Texas became a state is 41 times the two-digit number formed with its last two digits.

2. The year Texas became a state is 102.5 times the sum of all the digits of the year.

3. The second digit of the year that Texas became a state is one less than the sum of the third and the fourth (units) digits.

14. Wyoming became a state in 1890, after several years of being a territory. The number of years Wyoming spent as a territory (before becoming a state) is one less than the sum of all digits of the year when the territory was organized in the 19th century. When was the territory of Wyoming organized?

15. Wyoming became a state in 1890, after several years of being a territory. The number of years Wyoming spent as a territory (before becoming a state) is one less than the sum of all the digits of the year when the territory was organized in the 19th century. The units digit of the year is two more than the tens digit. When was the territory of Wyoming organized?

16. Ohio is one of the 50 US states. Its name came from Iroquois word "ohi-yo", meaning "great river" or "large creek". Although Ohio was admitted to the Union on March 1, 1803, the Congress forgot

to vote on a resolution that admits it to the Union until 1953. Therefore, Ohio received a statehood number different from what the state could have received in 1803. What are the Ohio official statehood and its "should have been" statehood numbers, if both numbers are prime numbers? Their average is higher than their difference. The difference has only three different factors in its prime factorization.

Answers

1. 9, 5
2. 56, 48, 7
3. 1787, 1959
4. 2, 21
5. 1863, 35, 41
6. 1789
7. 1789
8. 1812, 1821
9. 1, 1812, no
10. 1, 1, no, 101, 151, no, 1818, 1812
11. 1803, 1863, 1836, 1876, 1896, 1819, 1859, 1889
12. 1845
13. 2, 1845, yes
14. 1868
15. 1868
16. 47, 17

Solutions

1. Let u and p be the numbers of unpopulated and populated territories, then $\begin{cases} u^4 - p^4 = 5936 \\ u^2 - p^2 = 56 \end{cases}$. Noticing that $u^4 - p^4 = (u^2)^2 - (p^2)^2$ and using the formula for the square of a difference (see Appendix I-AF) and the second equation, the first equation can be simplified to

$$u^4 - p^4 = (u^2 - p^2)(u^2 + p^2) = 56(u^2 + p^2) = 5936.$$

Then, the system becomes $\begin{cases} u^2 + p^2 = 106 \\ u^2 - p^2 = 56 \end{cases} \Rightarrow \begin{cases} u^2 = 81 \\ p^2 = 25 \end{cases} \Rightarrow u = 9, p = 5.$

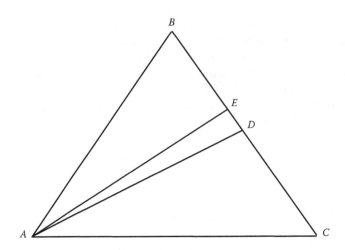

The United States of America has nine unpopulated and five populated territories in the Pacific and the Caribbean.

2. The area of a triangle is $A = \frac{base \cdot height}{2}$. Both $\triangle ABD$ and $\triangle ADC$ have the same area (Hint B) and the same height because there is only one line AE from the vertex A perpendicular to the side BC(see Figure I.1). Thus, $BD = CD$, and AD is a median. Then, using Hint A, we can write $BD = CD = AB/2$. As follows from Hint C, $AB + BD = 15$ or $AB + AB/2 = 15$, from which we can find $AB = 10$. Then $AC + DC = AC + AB/2 = 17$ leads to $AC = 12$.

1st way. The median from the vertex B in the isosceles triangle $\triangle ABC$ is the height to AC. The height is $\sqrt{10^2 - 6^2} = 8$. Then, the area of $\triangle ABC$ is $\frac{12 \cdot 8}{2} = 48$ sq. units.

2nd way. Using the Heron's formula (see Appendix I-T) the area is $A = \sqrt{16 \cdot (16 - 10)(16 - 12)(16 - 10)} = 48$.

The USA had 48 states (Hint A), 7 federal territories (Hint D), and the District of Columbia. Finally, $48 + 7 + 1 = 56$ in 1945.

The World War II Memorial in Washington has 56 pillars.

Alaska Territory, Territory of Hawaii, the Commonwealth of the Philippines, Puerto Rico, Guam, American Samoa, and Virgin Islands were seven federal territories of the United States of America in 1945.

3. The first digit is 1 (Hint E), though the problem can be solved without this hint. The sum of the other digits of the year when Delaware became a state is $2a + b = 22$ (Hints C and E), where a is

the second and fourth digits (Hint B) and b is the third. Hence, b is even and $b > 2$ to keep a as one digit. Hence, the options for Delaware are 1949, 1868, and 1787. Hints D and E for Hawaii lead to $2c + d = 23$, where c is the second and fourth digit (Hint B) and d is the third. Thus, d is odd and $d > 3$ to keep c as one digit. The options are 1959, 1878, and 1797.

Let us consider these options:

The year 1949 does not work for Delaware because then the only choice for Hawaii is 1959 (Hawaii became a state later) and 9 appears in both years that contradict the statement of the problem (Hint E).

1868 as a leap year does not work for Delaware (Hint A).

1787 for Delaware and 1959 for Hawaii satisfy the problem conditions.

Thus, Delaware became a state in 1787 and Hawaii became a state in 1959.

4. 1787 is a prime number. Thus, it has only two factors 1 and 1787. On the other side, 1818 can be presented as $1818 = 2^1 \cdot 3^2 \cdot 101^1$. It has (see Appendix I-PF) $(1 + 1) \cdot (2 + 1) \cdot (1 + 1) = 12$ factors, or 21 in the reverse order.

Pennsylvania became the 2nd state in 1787, while Illinois became the 21st state in 1818.

5. Let a be the two-digit number formed by the first two digits of the year, b and c be its third and fourth digits, and n and m be the current rank in size and the statehood number of West Virginia, then

$$\begin{cases} \dfrac{a}{b} = c \\ n - m = b \\ n + m = (a + 1)(c + 1) \end{cases}.$$

Since the sum of the first two digits is greater than 7, then a can be 17, 18, or 19. Two-digit numbers 17 and 19 do not work because they are prime numbers and cannot be a product of one-digit numbers b and c. Then $a = 18$. From $\frac{18}{b} = c$ follows that (b, c) can be from the set $\{(2, 9), (3, 6), (9, 2), (6, 3)\}$. The first two options $(2, 9)$ and $(3, 6)$ do not work because then $n + m = 190$ or $n + m = 133$ but there are only 50 states. It is obvious that the sum and difference of n and m are odd or even at the same time. Then b and $19(c + 1)$ should be odd or even simultaneously that can occur if $b = 6$, $c = 3$. From $\begin{cases} n - m = 6 \\ n + m = 76 \end{cases} \Rightarrow n = 41, m = 35.$

West Virginia became the 35th state in 1863. Its current rank in size is 41.

6. Let us solve $xy - 3y + 6x - x^2 = 29$ in y:

$$y(x - 3) - (x - 3)^2 = 20 \Rightarrow y = \frac{(x-3)^2 + 20}{x-3}.$$

Since y is a positive one-digit integer, $0 \le y = \frac{(x-3)^2 + 20}{x-3} \le 9$ and

$$\begin{cases} \dfrac{(x-3)^2 + 20}{x-3} \ge 0 \\ \dfrac{(x-3)^2 + 20}{x-3} \le 9 \end{cases} \Rightarrow \begin{cases} \dfrac{x^2 - 6x + 29}{x-3} \ge 0 \\ \dfrac{(x-3)^2 + 20}{x-3} - 9 \le 0 \end{cases} \Rightarrow \begin{cases} \dfrac{x^2 - 6x + 29}{x-3} \ge 0 \\ \dfrac{(x-7)(x-8)}{x-3} \le 0 \end{cases}$$

$$\Rightarrow \begin{cases} x \in (3, \infty) \\ x \in (-\infty, 3) \cup [7, 8] \end{cases}$$

Since x is a one-digit integer, it can take values 7 and 8. Both $x = 7$ and $x = 8$ lead to the same value for $y = 9$. The numbers 7, 8, and 9 form an arithmetic sequence with 9 as the last digit.

North Carolina became one of the original 13 states in 1789.

7. Since x and y are consequent terms of an arithmetic sequence (see Appendix I-SS) with the common difference of 1, then $y - x = 1$ and $xy - 3y + 6x - x^2 = 29$ can be rewritten as $x(1 + x) - 3(1 + x) + 6x - x^2 = 29 \Rightarrow 4x = 32$. Then $x = 8$, $y = 9$, and the second digit of the year is 7.

North Carolina became one of the original 13 states in 1789.

8. The equation of a circle with the center at $(-4, 1)$ and radius 4 is $(x + 4)^2 + (y - 1)^2 = 4^2 \Rightarrow x^2 + 8x + 16 + y^2 - 2y + 1 = 16 \Rightarrow x^2 + y^2 + 8x - 2y + 1 = 0$. Comparing it with the original equation, we obtain $a = 8$, $b = 2$, $c = 1$.

The Missouri Territory was established in 1812 and Missouri became a state in 1821.

9. The first digit of the year is 1. Let a be the second digit, b be the third digit, and c be the third one, then $$\begin{cases} 1000 + 100a + 10b + c = 151(10b + c) \\ 1 + a + b + c = 10b + c \end{cases}.$$ The second equation can be rewritten as $a = 9b - 1$, that leads to the only option of $b = 1$, $a = 8$. After substituting these values to the first equation, we have

$150c = 300 \Rightarrow c = 2$ and the year is 1812.

Louisiana joined the Union in 1812.

10. Let a, b, and c be the second, third, and fourth digits, then
$$\begin{cases} 1000 + 100a + 10b + c = k(10b + c) \\ 1 + a + b + c = 10b + c \end{cases}.$$ The second equation $a = 9b - 1$
leads to $b = 1$, $a = 8$. Substituting these values to the first equation, we get $1810 - 10k = c(k - 1)$ or $c = \frac{1800}{k-1} - 10$. Since $0 \le c \le 9$, $0 \le \frac{1800}{k-1} - 10 < 10 \Rightarrow 90 < k - 1 \le 180$. Moreover, $\frac{1800}{k-1}$ should be a natural number, i.e., $k - 1$ is a factor of 1,800 between 90 (not inclusive) and 180 inclusive. The four options, 100, 120, 150, and 180, produce one-digit $c = 8, 5, 2$, and 0 correspondingly. Then k can be 101, 121, 151, or 181, which lead to the following years: 1818, 1815, 1812, 1810, but only 1818 and 1812 are divisible by 6.

Louisiana joined the Union in 1812, Illinois joined the Union in 1818.

11. Rewriting Hint statements for the third digit in a mathematical way (see Appendix I-SS), we obtain

    1. 0, 6: the third digit is a perfect number (6 is the only one-digit perfect number) or a number that can be divided by any number but cannot divide any number (0).

    2. 3, 7, 9: the third digit, which is odd, is among one-digit Lucas numbers (that are 1, 3, 4, 7) or perfect squares (that are 1, 4, 9), but not both.

    3. 1,2, 5, 8: the third digit is a one-digit Fibonacci number (that are 1, 2, 3, 5, 8), not divisible by 3.

    Let a, b, and c be the second, third, and fourth digits, then

    $$\frac{10 + a}{c} = |a - c| + 1 \Leftrightarrow \begin{cases} \dfrac{10 + a}{c} = c - a + 1, & a < c, \\ \dfrac{10 + a}{c} = a - c + 1, & a \ge c. \end{cases}$$

If $a = c$, then $\frac{10+a}{a} = 1 \Rightarrow$ no solution.

If $a > c$, then $\frac{10+a}{c} = a - c + 1 \Rightarrow a = \frac{c^2 - c + 10}{c - 1} = c + \frac{10}{c-1}$. Because a and c are one-digit natural numbers, 10 should be divisible by $c - 1$. So, $c = 3$ or $c = 6$. In both cases $a = 8$. Considering Hint 1 for $c = 3$, we can find the following years: 1803 – Ohio, 1863 – West Virginia.

Hint 2 for $c = 6$ leads to the years: 1836 – Arkansas, 1876 – Colorado, 1896 – Utah.

Analogously, if $a < c$, then $\frac{10+a}{c} = c - a + 1 \Rightarrow a = \frac{c^2 + c - 10}{c+1} = c - \frac{10}{c+1}$ and 10 should be divisible by $c + 1$, then the only choice is $c = 9$, $a = 8$.

Considering the only hint that is left, Hint 3, we can get the years: 1819 – Alabama, 1829, 1859 – Oregon, 1889 – North Dakota, South Dakota Washington, Montana.

The following states joined the Union:

> 1803 – Ohio, 1863 – West Virginia.
> 1836 – Arkansas, 1876 – Colorado, 1896 – Utah.
> 1819 – Alabama, 1859 – Oregon, 1889 – North Dakota, South Dakota, Washington, Montana.

12. Let $1000 + 100a + 10b + c$ be the year when Texas became a state. Then

$$\begin{cases} 1 + a = b + c \\ \frac{10b + c}{10 + a} = \frac{5}{2} \end{cases} \Rightarrow \begin{cases} c = 1 + a - b \\ 20b + 2c = 50 + 5a \end{cases}.$$ Substitution c from the first

equation to the second leads to $a = 6b - 18$. Thus, b cannot be 0, 1, or 2 because then $a < 0$. Moreover, b cannot be greater than 5 because then a will be a two-digit number. Thus, 3 and 4 are the only options for b. If $b = 3$, then $a = 0$ and c is negative, from the first equation. Thus, $b = 4$, $a = 8$, and $c = 5$.

Texas became a state in 1845.

13. Let $1000 + 100a + 10b + c$ be the year when Texas became a state.

Then $\begin{cases} 1 + a = b + c \\ \frac{10b + c}{10 + a} = \frac{5}{2} \end{cases}$. The system has two equations and three

variables. Let us consider Hints to get the third equation.

From Hint 1 follows that $\frac{1000 + 100a + 10b + c}{10b + c} = 41 \Rightarrow 50 + 5a = 20b + 2c$, which is the second equation of the system. So, Hint 1 does not provide any new information.

Hint 2 leads to $1000 + 100a + 10b + c = 102.5(1 + a + b + c) \Rightarrow 5a + 185b + 203c = 1795$, which can be taken as the third equation.

Hint 3 produces $1 + a = b + c$, which is as the first equation. Therefore, Hint 3 does not provide any new information either.

Adding the equation of Hint 2, the initial system of equations becomes

$$\begin{cases} 1 + a = b + c \\ \dfrac{10b + c}{10 + a} = \dfrac{5}{2} \\ 1000 + 100a + 10b + c = 102.5(1 + a + b + c) \end{cases} \Rightarrow \begin{cases} c = 1 + a - b \\ 20b + 2c = 50 + 5a \\ 5a + 185b + 203c = 1795 \end{cases} \Rightarrow \begin{cases} c = 5 \\ a = 8. \\ b = 4 \end{cases}$$

Texas became a state in 1845.

Remark. The problem can be solved without using any hints. See solution to the previous problem.

Let us discuss three different ways to solve our system of three linear equations in three variables

$$\begin{cases} c = 1 + a - b \\ 20b + 2c = 50 + 5a \\ 5a + 185b + 203c = 1795 \end{cases} . \text{ It can be rewritten as } \begin{cases} a - b - c = -1 \\ -5a + 20b + 2c = 50 \\ 5a + 185b + 203c = 1795 \end{cases} .$$

The substitution method. Substituting $c = 1 + a - b$ from the first equation to the second and third equations reduces the system of three equations in three variables to the system of two equations in two variables:

$$\begin{cases} -3a + 18b = 48 \\ 208a - 18b = 1592 \end{cases} \Rightarrow \begin{cases} a = 6b - 16 \\ 104a - 9b = 796 \end{cases}. \text{ Then substitution of } a = 8b$$

-16 from the first equation to the second equation leads to $b = 4$. Then, $a = 8$ and $c = 5$.

The Gauss-Jordan elimination. Let us rewrite our system in a matrix form and perform operations on rows:

$$\begin{bmatrix} 1 & -1 & -1 & | & -1 \\ -5 & 20 & 2 & | & 50 \\ 5 & 185 & 203 & | & 1795 \end{bmatrix} \underset{\substack{II + 5 \cdot I \\ III + (-5) \cdot I}}{\Rightarrow} \begin{bmatrix} 1 & -1 & -1 & | & -1 \\ 0 & 15 & -3 & | & 45 \\ 0 & 190 & 208 & | & 1800 \end{bmatrix}$$

$$\underset{\substack{II' = II \cdot \frac{1}{15} \\ III + (-190) \cdot II'}}{\Rightarrow} \begin{bmatrix} 1 & -1 & -1 & | & -1 \\ 0 & 1 & -\frac{1}{5} & | & 3 \\ 0 & 0 & 246 & | & 1230 \end{bmatrix} \underset{\substack{III' = II \cdot \frac{1}{246} \\ II' = II + \frac{1}{5} \cdot III' \\ I - 1 \cdot III' + II'}}{\Rightarrow}$$

$$\begin{bmatrix} 1 & 0 & 0 & | & 8 \\ 0 & 1 & 0 & | & 4 \\ 0 & 0 & 1 & | & 5 \end{bmatrix} . \text{ Then, } a = 8, b = 4, \text{ and } c = 5.$$

The Cramer's rule requires calculating four determinants.

$$\Delta = \det\begin{bmatrix} 1 & -1 & -1 \\ -5 & 20 & 2 \\ 5 & 185 & 203 \end{bmatrix} = (1 \cdot 20 \cdot 203) + (-5 \cdot 185 \cdot (-1)) + (-1 \cdot 2 \cdot 5)$$

$$- (-1 \cdot 20 \cdot 5) - (-1 \cdot 2 \cdot 185) - (-5 \cdot (-1) \cdot 203) = 3690$$

$$\Delta_a \det\begin{bmatrix} -1 & -1 & -1 \\ 50 & 20 & 2 \\ 1795 & 185 & 203 \end{bmatrix} = 29520 \Rightarrow a = \frac{\Delta_a}{\Delta} = 8; \; \Delta_b = \det\begin{bmatrix} 1 & 11 & -1 \\ -5 & 50 & 2 \\ 5 & 1795 & 203 \end{bmatrix}$$

$$= 14760 \Rightarrow b = \frac{\Delta_b}{\Delta} = 4; \; \Delta_c = \det\begin{bmatrix} 1 & -1 & -1 \\ -5 & 20 & 50 \\ 5 & 185 & 1795 \end{bmatrix}$$

$$= 18450 \Rightarrow c = \frac{\Delta_c}{\Delta} = 5.$$

14. Let $1800 + 10a + b$ be the year when the territory of Wyoming was organized, then $(1 + 8 + a + b)$ is the sum of its digits, and $1890 = 1800 + 10a + b + (1 + 8 + a + b) - 1 \Rightarrow 82 = 11a + 2b$. Since b is a one-digit positive integer, from the last equation

$$0 \le b = \frac{82 - 11a}{2} \le 9 \Rightarrow \begin{cases} 82 - 11a \ge 0 \\ 82 - 11a \le 18 \end{cases} \Rightarrow \begin{cases} a \le 7.45 \\ a \ge 5.82 \end{cases}$$

Only two positive integers 6 and 7 satisfy the last system. When $a = 7 \Rightarrow b = 2.5$, which does not work. When $a = 6 \Rightarrow b = 8$.

The territory of Wyoming was organized in 1868.

15. Let a and b represent the tens and units digits correspondingly, then $a = b - 2$, and $1800 + 10a + b = 1800 + 11b - 20$ is the year when the Territory of Wyoming was organized. The sum of the digits of the year is $(1 + 8 + a + b) = (7 + 2b)$. The year when the Territory of Wyoming was organized can be presented via the year when Wyoming became a state: $1800 + 11b - 20 = 1890 - (7 + 2b - 1)$. The solution to the last equation is $b = 8$, and then $a = 6$.

The territory of Wyoming was organized in 1868.

16. There are 50 states in the US. Only two numbers less than or equal to 50 have three different factors in its prime factorization: $30 = 2 \cdot 3 \cdot 5$ and $42 = 2 \cdot 3 \cdot 7$. A set of primes below 50, their difference of either 42 or 30, and their average includes {(5, 47, 42, 26), (7, 37, 30,

22), (11, 41, 30, 26), (17, 47, 30, 32)}. The last pair (17, 47) has the mean higher than the difference between them.

Ohio became the 47th, and not the 17th state.

I.2 Presidents of The United States of America

People of the USA elect the president who represents them and their country. Each new elected president takes the oath of office and gives the inauguration address.

1. The $\dfrac{\sqrt{28+6\sqrt{3}}}{\sqrt{7-4\sqrt{3}}} - \dfrac{\sqrt{28-6\sqrt{3}}}{\sqrt{7+4\sqrt{3}}}$ Amendment to the US Constitution limits

   the number of terms a person can be elected to be president to two, whether consecutive or not, and the total years served as the president to ten. What amendment does set up presidential tenure? Evaluate without a calculator.

2. The first President of the United States, George Washington, informally limited the number of presidential terms by two refusing to run for the third term. Franklin D. Roosevelt was the first and the only American President to break George Washington's tradition. He died in office a few months after starting his new term.

   The order that Franklin D. Roosevelt has in the sequence of American Presidents and the number of times Franklin D. Roosevelt was inaugurated as the President of the United States are the largest and smallest values that satisfy

   $$\log_2 x^2 - 2^2 = 2{\cdot}5 - 2^2{\cdot}\log_x 2^5.$$

   Where does Franklin D. Roosevelt fall in the sequence of US presidents? How many terms did he serve as the US president?

3. Stephen Grover Cleveland (1837–1908) is the only President in the American history to serve two non-consecutive terms in office. Moreover, he won the popular vote three times, in 1884, 1888, and 1892.

   Stephen Grover Cleveland was the x-th and the y-th president of the USA where x and y satisfy the following system

   $$\begin{cases} y^6 - x^6 = 77723072 \\ y^3 - x^3 = 3176 \end{cases}$$

What are Grover Cleveland's numbers among the US presidents?

4. There are two cases in the US history when a father and later his son were elected the President of the United States: John Adams and John Quincy Adams and George H. W. Bush and George Walker Bush. In both cases during the election of the sons, the United States House of Representatives had to decide the winner following the Twelfth Amendment to the United States Constitution.

   Determine the number in the sequence of American presidents that the fathers, President John Adams and President George H. W. Bush, and their sons, President John Quincy Adams and President George Walker Bush, have. Use the hints below.

   1. President George H. W. Bush was the n-th president of the United States where the value n coincides with the length of the smallest leg of a right triangle with the hypotenuse of 841 units.

   2. The number of the presidency of his son, President George W. Bush, coincides with the length of the smallest leg of a new right triangle. The longer leg and hypotenuse of this new triangle have been increased by 84 compared to the longer leg and hypotenuse in the first triangle of Hint 1. All sides of both triangles are integers.

   3. Each digit in 84 represents the power of one number that corresponds to the number in the sequence of American presidents that President John Adams had. The number of his son's presidency is the product of the related exponents.

5. The Nobel Prize is an international award for outstanding achievements in the fields of physics, chemistry, medicine, literature, economic science, and peace. The prize ceremony takes place annually on December 10. Many outstanding scientists, inventors, and political leaders have received this prestigious award since it was established in 1901 following the will of Swedish inventor Albert Nobel (1833–1896).

   There are several Nobel Prize recipients from the White House. The number of presidents who won the Nobel Peace Prize after they left the office is the number of roots of the equation

   $$(x + 1)^2 + (x + 1)^{\frac{1}{2}} = 18.$$

   The largest root of the equation stands for the number of sitting US presidents when awarded the Nobel Peace Prize.

The number of US vice presidents who won this prestigious award while in office coincides with the number of positive solutions of the equation.

The number of US vice presidents who won the Nobel Prize after leaving the office coincides with the number of negative solutions of the equation.

How many US presidents (after and while in office) and US vice presidents (after and while in office) have been awarded the Nobel Prize? Can you name them?

6. The first President of the United States, George Washington, gave his inauguration speech on March 4 of a four-digit year. The third digit (tens) of the year is the square root of the fourth digit (ones). The cube root of the sum of the first and second digits is one less than the fourth digit. The fourth root of the sum of the second and third digits is the difference between the fourth and first digits. When did George Washington give his inauguration speech?

   Remark. It is obvious that the first digit is 1. Solve this problem without knowing this fact.

7. The first President of the United States, George Washington, gave his inauguration speech on March 4, 1793. How many possible numbers can be formed from the digits of the year? Is 1793 a prime number? List prime numbers that can be formed from the digits of the year.

8. The first US President, George Washington, gave the shortest inauguration speech in American history on March 4, 1793. How many words did the inauguration speech have if this number is the largest product of integers x and y that satisfy the equation

$$3y(1y - 3x - 6) - 3x(1x^2 + 2x + 3) = 46 - x^2(y - 1)?$$

9. The ninth President of the United States, William Henry Harrison, was the *last* president to be born as an English subject. He served the *shortest* term as the US president. Giving the *longest* inaugural address for an hour and 45 minutes, President Harrison caught a cold and died from pneumonia. His *grandson*, Benjamin Harrison, served a full term as the 23rd US president. It has been the only *grandfather-grandson* president duo in the US history.

   How many years did President Harrison serve as the President of the United States if this number is the product xy, where x and y satisfy the following system

$$\begin{cases} \dfrac{1}{x+y} = \dfrac{12}{7} \\[2mm] \dfrac{1}{x^2+y^2} = \dfrac{144}{25} \end{cases}?$$

Can you find the product xy without finding x and y first?

10. Virginia is the birthplace of more US presidents than any other state. The number of presidents born in Virginia is equal to the value of b such that the quadratic equation

$$x^2 + (b - 24)x + 5b + 24 = 0$$

has one solution (i.e., $x_1 = x_2$). Moreover, this solution is related to b in a surprising way and is not less than b.

How many US presidents were born in Virginia? Can you name them?

11. Virginia and Ohio are the birthplaces of the most US presidents. The numbers of US presidents that were born in Virginia and Ohio coincide with x and y correspondingly, $x > y$, that satisfy the equation

$$x^4 - y^4 - 36x^2 + 8y^2 + 217 = 0.$$

How many US presidents were born in Virginia and Ohio? Can you name them?

12. The $(q/2)$-th president of the United States is the only president to have founded an institution of higher education, University of Virginia, in 1819 on the land once belonged to the p-th president of the USA.

After finishing his undergraduate degree in 2 years, the $(q/3)$-th president of the USA stayed at the university for an additional year and became the Ivy League institution's first graduate student.

The $(p - 1)(q + 1)$-th president of the USA is the only president in the US history to hold a PhD. He received a doctorate degree in Political Science and History from Johns Hopkins University and passed the Georgia Bar Exam without graduating from a law school.

Name the presidents and give their number in the list of US presidents if p and q are the smallest integers such that four quadratic equations

$$x^2 + px + q = 0,$$
$$x^2 - px + q = 0,$$
$$x^2 + px - q = 0,$$
$$x^2 - px - q = 0,$$

have integer solutions.

13. "OK" or "Okay" is used to express approval or acknowledgment. There are many stories about its origin. One of them is connected to US President Martin Van Buren (1782–1862). He was from Kinderhook, NY, which was also called "Old Kinderhook". His supporters were known as "O.K. Clubs". Van Buren became the first president who was born after the American Revolution though English was his second language after Dutch. Van Buren also served as the governor of New York.

    What were the numbers of his presidency and governorship if these numbers are the numerator and denominator of the sum of

$$1 - \frac{1}{8} + \frac{1}{64} - \frac{1}{512} + \frac{1}{4096} - ....?$$

14. How many vice presidents have become US presidents if this number coincides with the x-intercept of $3y + 4x = 60$?

15. The third digit of the year when Air Force One, the official plane of the US president, was built is the value of k for which all solutions to the quadratic equation

$$(k - 3)x^2 + 2(k - 11)x + 3k + 3 = 0$$

    are equal. The solution to the equation is the last digit of the year. What year in the 20th century was the US Air Force One built?

Answers

1. 22
2. 32nd, 4 times
3. 22nd and 24th
4. 2, 41, 3, 43
5. 1, 3, 1, 0

6. 1793

7. 64, prime, 3, 7, 13, 17, 19, 31, 37, 71, 137, 139, ...

8. 133

9. 1/12 of a year or 31 days, yes

10. 8

11. 8, 7

12. 2, 3, 5, 28

13. 8, 9

14. 15

15. 1953

Solutions

1. Noticing that $28 = 1 + 27 = 1^2 + (3\sqrt{3})^2$, $6\sqrt{3} = 2 \cdot 1 \cdot 2\sqrt{3}$, $7 = 4 + 3 = 2^2 + (\sqrt{3})^2$, and $4\sqrt{3} = 2 \cdot 2 \cdot \sqrt{3}$, and applying formulas for the square of a sum or a difference, we obtain

$$\frac{\sqrt{28 + 6\sqrt{3}}}{\sqrt{7 - 4\sqrt{3}}} - \frac{\sqrt{28 - 6\sqrt{3}}}{\sqrt{7 + 4\sqrt{3}}} = \frac{\sqrt{1 + 2 \cdot 1 \cdot 3\sqrt{3} + 27}}{\sqrt{4 - 2 \cdot 2 \cdot \sqrt{3} + 3}} - \frac{\sqrt{1^2 - 2 \cdot 1 \cdot 3\sqrt{3} + (3\sqrt{3})^2}}{\sqrt{2^2 + 2 \cdot 2 \cdot \sqrt{3} + (\sqrt{3})^2}}$$

$$= \frac{\sqrt{(1 + 3\sqrt{3})^2}}{\sqrt{(2 - \sqrt{3})^2}} - \frac{\sqrt{(1 - 3\sqrt{3})^2}}{\sqrt{(2 + \sqrt{3})^2}}$$

$$= \frac{1 + 3\sqrt{3}}{2 - \sqrt{3}} - \frac{3\sqrt{3} - 1}{2 + \sqrt{3}}$$

$$= \frac{(1 + 3\sqrt{3})(2 + \sqrt{3}) - (3\sqrt{3} - 1)(2 - \sqrt{3})}{(2 - \sqrt{3})(2 + \sqrt{3})}$$

$$= \frac{22}{1}.$$

A detailed explanation of the solution technique is provided in Appendix I-AF.

The Twenty-Second Amendment describes the presidential tenure.

2. Let us simplify the equation $\log_2 x^2 - 2^2 = 2 \cdot 5 - 2^2 \cdot \log_x 2^5$ using properties of logarithms (see Appendix I-EL): $2 \cdot \log_2 x - 2^2 = 2 \cdot 5 - 2^2 \cdot 5 \cdot \log_x 2 \Rightarrow \log_2 x - 2 = 5 - 10\log_x 2 \Rightarrow \log_2 x - 7 + \frac{10}{\log_2 x} = 0$.

Multiplying both sides of the equation by $\log_2 x$, we obtain the quadratic equation with respect to $\log_2 x$: $\log_2^2 x - 7\log_2 x +$

$10 = 0 \Rightarrow (\log_2 x - 2)(\log_2 x - 5) = 0 \Rightarrow \log_2 x = 2$ or $\log_2 x = 5 \Rightarrow x = 4$ or $x = 32$.

Remark. The substitution $t = \log_2 x$ can be introduced to simplify the solution. Then the quadratic equation becomes: $t^2 - 7t + 10 = 0$.

Franklin D. Roosevelt was the 32nd President of the United States US president and he passed away a few months after starting his fourth term.

3. Using the formulas for the difference of cubes and squares (see Appendix I-AF), we obtain $\begin{cases} y^6 - x^6 = 77,723,072 \\ y^3 - x^3 = 3176 \end{cases}$

$\Rightarrow \begin{cases} (y^3)^2 - (x^3)^2 = 77,723,072 \\ y^3 - x^3 = 3176 \end{cases} \Rightarrow \begin{cases} (y^3 - x^3)(y^3 + x^3) = 77,723,072 \\ y^3 - x^3 = 3176 \end{cases}$.

Substituting the second equation to the first one and simplifying it, we get $\begin{cases} y^3 + x^3 = 24,472 \\ y^3 - x^3 = 3176 \end{cases}$. Adding two equations leads to $2y^3 = 27,648 \Rightarrow y^3 = 13,824 \Rightarrow y = 24$. Then $x = 22$.

Mr. Grover Cleveland was the 22nd and 24th US President.

4. **Hint 1:** The integer values of sides of a right triangle are Pythagorean triples (see Appendix I-PT). Then the value 841 of the hypotenuse can be presented by $\begin{cases} \frac{n^2}{4} + 1; & n - even \\ \frac{n^2 + 1}{2}; & n - odd \end{cases}$, where n is the smallest number (or the smallest side of the triangle). In the first case $\frac{n^2}{4} + 1 = 841 \Rightarrow n^2 = 3360$, which does not lead to an integer n, while $\frac{n^2 + 1}{2} = 841$ gives $n = 41$.

Hint 2: In the right triangle with the hypotenuse of 841 and one leg of 41, the second side is $\sqrt{841^2 - 41^2} = 840$. The second side can be also calculated using $\frac{n^2 - 1}{2} = 840$. Then, the second triangle has the hypotenuse of $841 + 84 = 925$, the longest leg of $840 + 84 = 924$, and the smallest leg of $\sqrt{925^2 - 924^2} = 43$.

Hints 3: Digits of 84 are 8 and 4, which are the third and second power of 2. Finally, the product of exponents $3 \cdot 2 = 6$.

John Adams was the second President of the United States, John Quincy Adams was the sixth President, George H. W. Bush was the 41st President, and George W. Bush was the 43rd President of the United States

5. The domain of the equation $(x + 1)^2 + (x + 1)^{\frac{1}{2}} = 18$ is $[-1, \infty)$.

Let us consider the interval $[0, \infty)$ first. On this interval the function $y = (x + 1)^2 + (x + 1)^{\frac{1}{2}}$ is strictly increasing. In addition $y(0) = -16$ and $y \to \infty$ as $x \to \infty$. Thus, the equation $(x + 1)^2 + (x + 1)^{\frac{1}{2}} = 18$ has just one solution. It is easy to see that this solution is 3.

On the interval $[-1, 0)$, the term $(x + 1)^2$ takes values between 0 and 1 and $(x + 1)^{\frac{1}{2}}$ between 0 and $\sqrt{2}$. Thus, their sum cannot be 18, and there is no solution on $[-1, 0)$. There are no negative solutions.

US presidents while in office:

Theodore Roosevelt (1858–1919, in office 1901–1909) was awarded the Nobel Prize in 1906 "for his successful mediation to end the Russo-Japanese war and for his interest in arbitration, having provided the Hague arbitration court with its very first case".

Woodrow Wilson (1856–1924, in office 1913–1921) was awarded the Nobel Prize in 1919 for founding the League of Nations, the predecessor to the United Nations.

Barack Obama (born 1961, in office 2009–2016) was awarded the Nobel Prize in 2009 "for his extraordinary efforts to strengthen international diplomacy and cooperation between peoples".

US President after leaving the office:

Jimmy Carter (born in 1924, in office 1977–1981) was awarded the Nobel Prize in 2002 "for his decades of untiring effort to find peaceful solutions to international conflicts, to advance democracy and human rights, and to promote economic and social development".

US Vice President:

Al Gore (born in 1948, in office 1993–2001) was awarded the Nobel Prize in 2007 "for his efforts to obtain and spread knowledge about climate change".

6. Let a, b, c, and d be the first, second, third, and fourth (units) digits

of the year. Then $\begin{cases} \sqrt{c} = d \\ \sqrt[3]{a + b} = d - 1. \\ \sqrt[4]{b + c} = d - a \end{cases}$

Because all the digits are positive integers between 0 and 9 and $a \neq 0$, the first equation leads to the set $(c, d) \in \{(0, 0), (1, 1), (4, 2), (9, 3)\}$.

The second equation gives $(a + b, d) \in \{(1, 2), (8, 3)\}$, which immediately rules out $\{(0, 0), (1, 1)\}$ from the set of (c, d). So, $d = 2$ or $d = 3$.

If $d = 2$, $c = 4$, then $a + b = 1$ and the third equations leads to $5 - a = (2 - a)^4$ that does not have an integer solution.

If $d = 3$, $c = 9$, then $a + b = 8$ and the third equation is $\sqrt[4]{b + 9} = 3 - a$ or $\sqrt[4]{17 - a} = 3 - a \Rightarrow a = 1$, $b = 7$.

US President George Washington gave his inauguration speech in 1793.

7. The number of drawing n elements from a set of N elements and arranging them in distinct order is determined by the permutation $P_n^N = \frac{N!}{(N-n)!}$ (see Appendix I-CR). Then the numbers of one-, two-, three-, and four-digit numbers are $P_1^4 = \frac{4!}{(4-1)!} = 4$, $P_2^4 = \frac{4!}{(4-2)!} = 12$, $P_3^4 = \frac{4!}{(4-3)!} = 24$, $P_4^4 = \frac{4!}{(4-4)!} = 24$. Thus, $4 + 12 + 24 + 24 = 64$ distinct numbers can be formed from the digits of 1793. Some of prime numbers are 3, 7, 17, 19, and so on.

8. Let us rearrange the terms of the equation $3y(1y - 3x - 6) - 3x(1x^2 + 2x + 3) = 46 - x^2(y - 1)$ as $-3x^3 + x^2y - 7x^2 + 3y^2 - 9xy - 18y - 9x = 46 \Rightarrow -3x(x^2 + 3y + 3) + y(x^2 + 3y + 3) - 7(x^2 + 3y + 3) = 25 \Rightarrow (x^2 + 3y + 3)(y - 3x - 7) = 25$. Since x and y are integers and 25 has factors $\{-1, 1, -5, 5, -25, 25\}$, we can obtain 6 systems of equations:

$$\begin{cases} x^2 + 3y + 3 = 25 \\ y - 3x - 7 = 1 \end{cases}; \begin{cases} x^2 + 3y + 3 = 1 \\ y - 3x - 7 = 25 \end{cases}; \begin{cases} x^2 + 3y + 3 = -25 \\ y - 3x - 7 = -1 \end{cases};$$

$$\begin{cases} x^2 + 3y + 3 = -1 \\ y - 3x - 7 = -25 \end{cases}; \begin{cases} x^2 + 3y + 3 = 5 \\ y - 3x - 7 = 5 \end{cases}; \begin{cases} x^2 + 3y + 3 = -5 \\ y - 3x - 7 = -5 \end{cases}.$$

The second, third, and fifth systems do not have real solutions. The first and fourth systems do not have integer solutions. The sixth system has:

$$\begin{cases} x^2 + 3y + 3 = -5 \\ y - 3x - 7 = -5 \end{cases} \Rightarrow \begin{cases} x^2 + 9x + 14 = 0 \\ y = 3x + 2 \end{cases} \Rightarrow \begin{matrix} x = -7, y = -19 \\ x = -2, y = -4 \end{matrix}.$$

The inauguration speech that George Washington gave in 1793 had just $(-7)(-19) = 133$ words.

9. Let us rewrite our system $\begin{cases} \dfrac{1}{x+y} = \dfrac{12}{7} \\ \dfrac{1}{x^2+y^2} = \dfrac{144}{25} \end{cases}$ as $\begin{cases} x + y = \dfrac{7}{12} \\ x^2 + y^2 = \dfrac{25}{144} \end{cases}$. Raising both sides of the first equation to the second power

$(x + y)^2 = \frac{49}{144} \Rightarrow x^2 + 2xy + y^2 = \frac{49}{144}$ that after substituting to

$x^2 + y^2 = \frac{25}{144}$ becomes $2xy = \frac{49}{144} - \frac{25}{144}$. Then $xy = 1/12$ or 31 days.

President Harrison served in the office only for one month.

10. Two repeated solutions to a quadratic equation appear if a discriminant is zero (see Appendix I-QE). The discriminant of the quadratic equation $x^2 + (b - 24)x + 5b + 24 = 0$ is $D = (b - 24)^2 - 4(5b + 24)$. It is zero if $b^2 - 68b + 480 = 0 \Rightarrow b = 8$ or $b = 60$.

If $b = 8$ then $x = 8$ $(x \geq b)$. If $b = 60$, then $x = 18$ $(x < b)$.

Eight US presidents were born in Virginia: George Washington, Thomas Jefferson, James Madison, James Monroe, William Henry Harrison, John Tyler, Zachary Taylor, and Woodrow Wilson.

11. Let us rearrange and combine terms as $x^4 - y^4 - 36x^2 + 8y^2 + 217 = 0 \Rightarrow$ $(x^4 - 2 \cdot 18x^2 + 18^2) - (y^4 - 2 \cdot 4y^2 + 4^2) + 217 - 18^2 + 4^2 = 0 \Rightarrow (x^2 - 18)^2$ $- (y^2 - 4)^2 = 91 \Rightarrow ((x^2 - 18) - (y^2 - 4))((x^2 - 18) + (y^2 - 4)) = 91$. The variables x and y are integers, $x > y$, and 91 is prime. Therefore, the only possible choices are

$$\begin{cases} x^2 - y^2 - 14 = 1 \\ x^2 + y^2 - 22 = 91 \end{cases}, \quad \begin{cases} x^2 - y^2 - 14 = 91 \\ x^2 + y^2 - 22 = 1 \end{cases}, \quad \begin{cases} x^2 - y^2 - 14 = -1 \\ x^2 + y^2 - 22 = -91 \end{cases},$$

and $\begin{cases} x^2 - y^2 - 14 = -91 \\ x^2 + y^2 - 22 = -1 \end{cases}$.

From the first system we get $\begin{cases} x^2 - y^2 = 15 \\ x^2 + y^2 = 113 \end{cases} \Rightarrow \begin{cases} x^2 = 64 \\ y^2 = 49 \end{cases} \Rightarrow \begin{cases} x = \pm 8 \\ y = \pm 7 \end{cases}$.

The second system $\begin{cases} x^2 - y^2 = 105 \\ x^2 + y^2 = 23 \end{cases}$ does not have a solution because

the sum of squares cannot be less than their difference.

The third system $\begin{cases} x^2 - y^2 = 13 \\ x^2 + y^2 = -69 \end{cases}$ does not have a solution because

the sum of two squares cannot be negative.

The first equation $x^2 - y^2 = -77$ of the fourth system does not satisfy the problem because x should not be less than y.

Eight US presidents were born in Virginia: William Henry Harrison, Thomas Jefferson, James Madison, James Monroe, John

Tyler, Zachary Taylor, George Washington, and Woodrow Wilson.

Seven US presidents were born in Ohio: Ulysses Simpson Grant, Rutherford Birchard Hayes, James Abram Garfield, Benjamin Harrison, William McKinley, William Howard Taft, and Warren Gamaliel Harding.

12. The expressions p, q, $q/2$ and $q/3$ represent counting numbers. Thus, they are positive integers and q should be divisible by $6 = 2 \cdot 3$.

Solutions of the quadratic equations are integers if their discriminants $(p^2 - 4q)$ and $(p^2 + 4q)$ are squares. The smallest proper integer $q = 6$ leads to $(p^2 - 4 \cdot 6)$ and $(p^2 + 4 \cdot 6)$ that should be squares. The smallest p that produces this outcome is $p = 5$.

Other ways to solve the problem can be based on considering $(p^2 - 4q) = m^2$ and $(p^2 + 4q) = n^2$ and $p \geq 3$ that follows from $(p^2 - 4q) > 0$.

The 2nd President James Madison, 3rd President Thomas Jefferson, 5th President James Monroe, 28th President Woodrow Wilson.

13. $1 - \frac{1}{8} + \frac{1}{64} - \frac{1}{512} + \frac{1}{4096} - \dots$ is a geometric series with the first term of 1 and the ratio of $-1/8$. Thus, $S = 1 \cdot \frac{1}{1 - \left(-\frac{1}{8}\right)} = \frac{8}{9}$ (see Appendix I-SS).

Martin Van Buren served 8 years as the US president and 9 years as the governor of New York.

14. The x-intercept occurs at the point where $y = 0$ or at $(15, 0)$. Fifteen US vice presidents became US presidents.

15. The roots of a quadratic equation $ax^2 + bx + c = 0$ are equal if its discriminant is zero (see Appendix I-QE), i.e., $b^2 - 4ac = 0 \Rightarrow (2(k - 11))^2 - 4(k - 3)(3k + 3) = 0 \Rightarrow k^2 + 8k - 65 = 0 \Rightarrow k = -13$ or $k = 5$.

$k = -13$ cannot be a digit of a year. So, the third digit of the year is 5. Finally, 3 satisfies $x^2 - 6x + 9 = 0$ obtained after substituting 5 into the original equation.

The US Air Force One was built in 1953.

I.3 The American Flag

A national flag symbolizes a country. The American flag has 13 horizontal red stripes alternating with white stripes and a blue rectangle with 50 white stars in the canton. The 50 five-pointed stars are arranged in nine horizontal rows with rows of six stars alternating with rows of five stars representing the 50 states of the United States of America. The 13 stripes symbolize the first 13 states in the Union that used to be British colonies and declared independence from the Kingdom of Great Britain.

Throughout the years, different designs of the flag have been called the Stars and Stripes, Red, White and Blue, Old Glory, and The Star-Spangled Banner.

The first 13-star American flag was adopted on June 14, 1777 and its current 50-star version was adopted on July 4, 1960.

1. The design of the current American flag is specified as

   Hoist (height) of the flag: $A = 1.0$;
   Fly (width) of the flag: $B = 1.9$;
   Hoist (height) of the canton: $C = \frac{7}{13}A$;
   Fly (width) of the canton: $D = \frac{2}{5}B$.

   What is the ratio between the hoist and fly of the canton?

2. The design of the current national flag of the United States of America has the following dimensions

   Hoist (height) of the flag: $A = 1.0$;
   Fly (width) of the flag: $B = 1.9$;
   Hoist (height) of the canton ("union"): $C = \frac{7}{13}A$;
   Fly (width) of the canton: $D = \frac{2}{5}B$.

   What portion of the flag does the canton take? Can the problem be solved without some data provided?

3. The 50 five-pointed stars (pentagrams) on the American flag represent the 50 US states.

   The diameter of a circumscribed circle of a five-pointed star is $2r$. What is the area of the star?

4. The American flag designed by Francis Hopkinson in 1777 had six-pointed stars. What is the area of a star if the diameter of a circumscribed circle of the star is $2r$?

5. The design of the American flag has been modified several times. Numbers of stripes on the *Star-Spangled Banner Flag* and on the earlier *Betsy Ross Flag* are different.

The numbers of stripes in the *Betsy Ross Flag* and the *Star-Spangled Banner Flag* are the values of a and b, respectively, when the system

$$\begin{cases} 4y = a + 3x \\ bx = 20y - 65 \end{cases} \text{ has infinitely many solutions.}$$

How many stripes are in the *Betsy Ross Flag* and the *Star-Spangled Banner Flag*? For what values of a and b does the system have no, infinitely many, or a unique solution?

6. The stars are arranged in a circle on the *Betsy Ross Flag*. The number of stars is a natural number in the domain of $\log_{x-11} \frac{12 + 13x}{14 - x}$. How many stars are on the *Betsy Ross Flag*?

7. The *Star-Spangled Banner Flag* inspired Francis Scott Key to write the poem *Defense of Fort McHenry*, which later became the national anthem of the United States.

   The number of stars on the *Star-Spangled Banner Flag* is a natural number in the domain of $\log_{x-13}(16 - x)$. How many stars are on the flag?

8. The *Star-Spangled Banner Flag* or the *Great Garrison Flag* was the garrison flag that flew over Fort McHenry in Baltimore Harbor during the War of 1812. It inspired Francis Scott Key to write the poem *Defense of Fort McHenry* in the 19th century, which was set to the 18th century tune *To Anacreon in Heaven* and became the national anthem of the United States in the 20th century. The last two digits of the years when the lyrics was written, the poem was adopted as the national anthem, and music was composed would be the first, fourth, and twelfth terms of an arithmetic sequence if the poem were adopted 1 year earlier. These three numbers would be the first, second, and fifth terms of the second arithmetic sequence if the music were composed 2 years earlier. The sum of the first four terms of the first arithmetic sequence is 1 less than the sum of the first three terms of the second arithmetic sequence.

   When were the lyrics and music of the national anthem of the United States written? When were they adopted as the national anthem?

9. The flag day is celebrated on June 14, the day when the Second Continental Congress passed the Flag Resolution. The first official US flag was flown during the Siege of Fort Stanwix at Fort Schuyler on August 3 later that year. Soldiers cut up their shirts to make the

white stripes and women made the red stripes from their red flannel petticoats.

What year did the Second Continental Congress pass the Flag Resolution if each digit of the year is a solution to the polynomial equation

$$x^4 - 22x^3 + 168x^2 - 490x + 343 = 0,$$

arranged in a nondecreasing order?

10. The US flag appeared on a US postal stamp in the 20th century depicting the flag with a circle of stars. The last two digits of this year form a number, which is twice the number of stars on the flags. The average of these two numbers (the number of stars and the last two-digit number) is greater than the number formed by the first two digits of the year. The lowest integer that satisfies these conditions leads the number of stars on the flag. What year did the first American flag stamp appear? How many stars were on the flag?

11. Each state in the USA has its state flag with a unique design and motto.

The smallest positive integer coefficients a and b in the following four quadratic equations ($a \neq 0$)

$$ax^2 + bx + 6 = 0,$$
$$ax^2 + bx - 6 = 0,$$
$$ax^2 - bx + 6 = 0,$$
$$ax^2 - bx - 6 = 0,$$

are such that two solutions of each equation are different integers. The smallest positive value from a group of all solutions that satisfy at least one equation gives the number of the US states with the state flag featuring the Union Flag of the UK. The largest solution is the number of flags that Texas has in its motto. The second smallest positive integer solution is the number of stripes in the North Carolina state flag. The third smallest positive integer solution is the number of stars in the Tennessee state flag as well as the number of stripes in the Colorado, Georgia, Mississippi, and Missouri state flags.

Describe each flag based on information provided.

Answers

1. 175/247
2. 14/65, yes
3. $1.12257r^2$
4. $\sqrt{3}\,r^2$
5. 13 and 15 stripes; no solutions if $a \neq 13$ and $b = 15$, a unique solution if $b \neq 15$.
6. 13
7. 15
8. 1814, 1780, 1931
9. 1777
10. 1926, 13
11. 1, 6, 2, 3

Solutions

1. $\dfrac{C}{D} = \dfrac{\frac{7}{13}A}{\frac{2}{5}B} = \dfrac{\frac{7}{13}A}{\frac{2}{5} \cdot \frac{19}{10}A} = \dfrac{175}{247}$.

   The hoist is related to the fly of the canton as 175 to 247.

2. $\dfrac{A_{canon}}{A_{flag}} = \dfrac{C \cdot D}{A \cdot B} = \dfrac{\frac{7}{13}A \cdot \frac{2}{5}B}{A \cdot B} = \dfrac{14}{65}$ is the portion of the flag that the canton takes. The actual values of the hoist A and the fly B of the flag are not used.

   The canton takes $\frac{14}{65}$ part of the flag.

3. In the triangle $\triangle AOC$: $\angle AOC = 360°/10 = 36°$, $\angle OAC = (360°/10)/2 = 18°$, and $\angle OCA = 180° - 18° - 36° = 126°$ (see Figure I.2).

   1st way: By Theorem of Sines (see Appendix I-T) $\Rightarrow \dfrac{r}{\sin 126°} = \dfrac{OC}{\sin 18°} = \dfrac{AC}{\sin 36°} \Rightarrow$

   by the Heron's formula $\Rightarrow A_{\triangle AOC} = \sqrt{p(p-r)\left(p - \dfrac{r\sin 18°}{\sin 126°}\right)\left(p - \dfrac{r\sin 36°}{\sin 126°}\right)} = $

   $0.112257r^2$, where p is a semi perimeter of $\triangle AOC$, $p = \dfrac{1}{2}\left(r + \dfrac{r\sin 18°}{\sin 126°} + \dfrac{r\sin 36°}{\sin 126°}\right)$.

   Then $A_{star} = 10\sqrt{p(p-r)\left(p - \dfrac{r\sin 18°}{\sin 126°}\right)\left(p - \dfrac{r\sin 36°}{\sin 126°}\right)} = 1.12257r^2$

   2nd way: By the Theorem of Sines $\Rightarrow \dfrac{r}{\sin 126°} = \dfrac{OC}{\sin 18°} \Rightarrow$ the height to AO in

   $\triangle AOC$, $h = OC \cdot \sin \angle AOC = \dfrac{r}{\sin 126°}\sin 18° \cdot \sin 36° \Rightarrow A_{\triangle AOC} = \dfrac{rh}{2} = \dfrac{r^2}{2\sin 126°}$

   $\sin 18° \cdot \sin 36°$ and $A_{star} = \dfrac{5r^2}{\sin 126°}\sin 18° \cdot \sin 36° = 1.12257r^2$

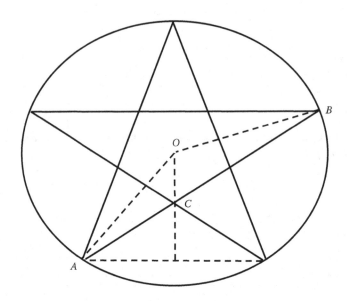

FIGURE I.2

3rd way: The area of $\triangle AOB$ is $A_{\triangle AOB} = r^2 \sin 18° \cos 18°$. Five triangles like $\triangle AOB$ form the star, but the inner pentagon is taken twice, its area $A_{pentagon} = 5r^2 \sin^2 18° \tan 36°$. Then $A_{star} = 5A_{\triangle AOC} - A_{pentagon} = 5 \cdot r^2 \sin 18°$ $\cos 18° - 5r^2 \sin^2 18° \tan 36° = 1.12257 r^2$.

There are other ways to solve the problem (Figure I.3).

4. In the triangle $\triangle AOC$, $\angle AOC = 360°/12 = 30°; \angle OAC = \frac{1}{2}\angle POQ = (360°/6)/2 = 30° \Rightarrow OC = CA$ and the height to OA in the $\triangle AOC$ is $h = \frac{r}{2} \cdot \tan \angle AOC = \frac{r\sqrt{3}}{6} \Rightarrow A = \frac{rh}{2} = \frac{r^2\sqrt{3}}{12}$ and $A_{star} = 12 \cdot \frac{r^2\sqrt{3}}{12} = \sqrt{3} r^2$.

5. The system $\begin{cases} 4y = a + 3x \\ bx = 20y - 65 \end{cases} \Rightarrow \begin{cases} 4y - 3x = a \\ -20y + bx = -65 \end{cases}$ has infinitely many solutions when the coefficients are proportional, i.e.,

$\frac{4}{-20} = \frac{-3}{b} = \frac{a}{-65}$, or $a = 13$, $b = 15$. The system does not have a solution if $a \neq 13$, $b = 15$. The system has a unique solution if $b \neq 15$.

The Betsy Ross Flag has 13 stripes, and the Star-Spangled Banner Flag has 15 stripes.

6. The domain (see Appendix I-EL) of $\log_{x-11} \frac{12 + 13x}{14 - x}$ is $\begin{cases} x - 11 > 0°, x - 11 \neq 1 \\ \frac{12 + 13x}{14 - x} > 0 \end{cases} \Rightarrow \begin{cases} x > 11, x \neq 12 \\ -\frac{12}{13} < x < 14 \end{cases}$. Thus, the only natural number in the domain is 13.

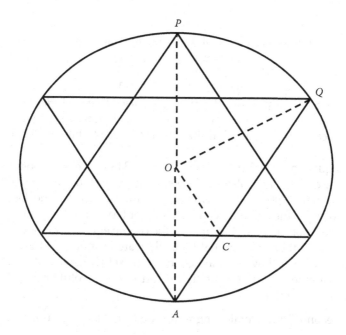

FIGURE I.3

The Betsy Ross Flag has 13 stars.

7. The domain (see Appendix I-EL) of $\log_{x-13}(16 - x)$ is
$$\begin{cases} x - 13 > °, x - 13 \neq 1 \\ 16 - x > ° \end{cases} \Rightarrow \begin{cases} x > 13, x \neq 14 \\ x < 16 \end{cases}.$$ Hence, the only natural number is 15.

The Star-Spangled Banner Flag has 15 stars.

8. The last two digits x of the year when lyrics was written is the first term of both arithmetic sequences. If d and b are differences of the first and second arithmetic sequences, then the last two digits of the years of adopting the national anthem and composing music are the fourth (without 1) and twelfth terms of the first arithmetic sequence, i.e., $(x + 3d - 1)$ and $(x + 11d)$. They are also the second and fifth (without 2) elements of the second arithmetic sequence, $(x + b)$ and $(x + 4b - 2)$. Thus, we obtain the following system
$$\begin{cases} x + 3d - 1 = x + b \\ x + 11d = x + 4b - 2 \\ 4x + 6d + 1 = 3x + 3b \end{cases}$$ with the last equation representing the sum of

three and four elements of the first and second sequences. The

system is reduced to $\begin{cases} 3d - 1 = b \\ 11d = 4b - 2 \\ x + 6d + 1 = 3b \end{cases} \Rightarrow \begin{cases} b = 17 \\ d = 6 \\ x = 14 \end{cases}$.

Francis Scott Key wrote the poem *Defense of Fort McHenry* in 1814 that was set to the music of *To Anacreon in Heaven* written in 1780 and became the national anthem of the United States in 1931.

9. The equation $x^4 - 22x^3 + 168x^2 - 490x + 343 = 0$ has 4 roots. It can have 4, 2, or 0 real roots (see Appendix I-PE). Because the signs of the coefficients of $p(-x)$, +−+−+, change 4 times, then by the Descartes Rule of Sign, the equation can have 4, 2, or 0 positive real roots and no negative roots. Indeed, the signs of the coefficients of $p(-x)$, +++++, are the same. If there are rational roots, they are factors of 343 divided by factors of 1, i.e., they are numbers from the set {1, 7, 49, 343}. Checking these numbers starting with 1, we can find the solutions: 1, 7, 7, 7, which lead to the year.

The Second Continental Congress passed the Flag Resolution in 1777.

10. The first two digits of the year in the 20th century are 19. Let x be the number of stars, then $2x$ is the last two digits of the year. Then from $(2x + x)/2 > 19$, we obtain $x = 13$ as the smallest integer that satisfy the inequality.

There are 13 stars on the first post stamp issued in 1926.

11. The solutions of the equations are integers and different if $b^2 - 4a \cdot 6$ and $b^2 + 4a \cdot 6$ are positive and squares (see Appendix I-QE), i.e., $b^2 - 4a \cdot 6 = m^2$ and $b^2 + 4a \cdot 6 = n^2$. Adding the last equations, we obtain $2b^2 = m^2 + n^2$. Moreover, from $b^2 - 4a \cdot 6 > 0$ follows that $b^2 > 4a \cdot 6$ or $b \geq 5$, i.e., $2b^2 \geq 50$ and $m^2 + n^2 \geq 50$. The smallest integer that b can take is 5 and let us test it. Then $m^2 + n^2 = 50$ and checking all squares less than 50 (1, 4, 16, 25, 36, 49) we can find the pair $m^2 = 1$, $n^2 = 49$, from which follows that $a = 1$. Then the solution to the first equation is (−2, −3), to the second is (1, −6), to the third is (2, 3), and to the fourth is (−1, 6). The set of all numbers that satisfy at least one equation is {−6, −3, −2, −1, 1, 2, 3, 6}.

The state flag of Hawaii features the Union Flag of the United Kingdom. Texas has 6 flags in its motto. North Carolina has 2 stripes on its state flag. Tennessee has 3 stars on its state flag. Colorado, Georgia, Mississippi, and Missouri have 3 stripes on their state flags.

I.4 Symbols of the US States

Each US state has its capital, flag, seal, and other state symbols such as a state tree, a state flower, and a state bird. The principle of state's rights is added to the United States Constitution under the 10th Amendment.

1. The Constitution of the United States of America was adopted on September 17, 1787. A total of 27 amendments have been added to the Constitution. The first amendment, named the Bill of Rights, guarantees fundamental civil rights and freedoms. They were ratified in 1791. How many amendments are in the Bill of Rights if the sum of cubes of amendments in the Bill of Rights and other amendments (added to the Constitution after the Bill of Rights, which is greater) is 5913?

2. Before adoption of standard time zones, many towns and cities in the continental US set time according to sunsets and sunrises at their locations. Over 300 local sun times brought confusion to transportation schedule and telecommunication systems in the 19th century.

   The continental United States was divided into

$$\frac{16}{1\cdot2\cdot3} + \frac{16}{2\cdot3\cdot4} + \frac{16}{3\cdot4\cdot5} + \frac{16}{4\cdot5\cdot6} + \frac{16}{5\cdot6\cdot7} + ...+ \frac{16}{n\cdot(n+1)\cdot(n+2)} + ...$$

   time zones on November 18, 1883. How many time zones are in the continental United States? How many time zones do the USA and its territories have today if this number is one-fifth of the reciprocal of the eighth term of the sequence?

3. The names of x states have the longest commonly used names that consist of z letters. Other x states have the shortest names with just y letters. The longest official state name has $7x$ letters. The variables x, y, and z are connected by

$$x + \frac{1}{y + \frac{1}{z}} = \frac{172}{53}.$$

   How many letters are in the shortest and longest commonly used state names? How many letters are in the longest official state name? Can you name these states?

4. Each state has adopted a state bird as one of its symbols. Sometimes the same bird has been chosen by several states.

   The number of states with a chickadee as their state bird is related to the number of states with a mockingbird as their state bird as the inradius is related to the circumradius of circles inscribed in and circumscribed around a right triangle. The sides of the triangle are consecutive terms of an arithmetic series. How many states have adopted chickadee or mockingbird as their state bird if the product of these numbers is the lowest among all possible options? Can you name these states?

5. Each state has chosen its own state flower and state tree.

   The number of states that selected the same plant to be their state flower and state tree is p in the quadratic equation $x^2 - 5x + p = 0$. The sum of its roots cubed is 95, i.e., $x_1^3 + x_2^3 = 95$, where x_1 and x_2 are solutions to the quadratic equation. How many states have the same plant for their state flower and state tree? Can you name the states and plants?

6. Either an oak as the state tree or a rose as the state flower has been chosen by several states. These numbers are the maximum non-negative integer values of x and y, correspondingly, from all possible pairs (x, y) that satisfy the equation $xy - x - 3y = 0$. Either a sequoia or a mountain ash tree has been adopted as the state tree by the numbers of states, which are the minimum nonnegative integer values of x and y correspondingly. The median of all integers x and y from all pairs of nonnegative integer solutions (x, y) provide the number of states that have the word either "blossom" (for x) or "magnolia" (for y) in their state flower. How many states have an oak, sequoia, or mountain ash tree as their state tree and "blossom", magnolia, or rose as their state flower? List these states if any.

7. A pine tree is the most popular state tree adopted by the largest number of states compared to any other tree. There are more than 100 different species of a pine tree.

   How many states have a pine tree as its state tree if this number is defined by the value of

   $$\log_a(\sqrt{x^2 + 1} + x) \text{ if } \log_a(\sqrt{x^2 + 1} - x) = -10.$$

   Can you name these states?

8. Santa Fe is the oldest serving capital and the highest capital in the US.

   The number formed from the first two digits of the year when Santa Fe became the capital and the number formed from the last two digits of the year when New Mexico became a state are the largest and smallest legs of a right triangle with the area of 96 sq. units and perimeter of 48 units. The number formed from the last two digits of the year when Santa Fe became the capital are the half of the hypotenuse. The number composed of the first two digits of the year when New Mexico became a state are one unit less than the hypotenuse.

When did Santa Fe become the capital? When did New Mexico become a state?

Remark. Two digits to form a two-digit number are taken in the order they appear in the year.

9. Pensacola used to be the capital of the Spanish colonies of East Florida and West Florida. Therefore, the first session of Florida's Legislative Council as a territory of the United States was at Pensacola. Delegates from St. Augustine traveled 59 days by water to attend. The second session was in St. Augustine. Western delegates traveled 28 days around the peninsula. During this session, it was decided that future meetings should be held at a halfway between these two largest cities, at Tallahassee.

Tallahassee became the capital of Florida at the year that can be presented by joining two numbers, which are the smallest and middle numbers of the Pythagorean triples that are consecutive terms of an arithmetic sequence. The sum of the numbers has the smallest value and 7 among its digits.

When did Tallahassee become the capital of the Florida Territory?

10. Boston, Harford, Providence, and Concord are the state capitals of Massachusetts, Connecticut, Rhode Island, and New Hampshire, respectively.

The geometric mean of the distances from Boston to the other three capitals, Harford, Providence, and Concord, is 63 miles. The sum of the squares of these distances is 14,509 square miles, while the sum of the inverses of the distances is 0.05 1/mile. What is the average distance between Boston and three other state capitals? Find the distances between Boston and three state capitals, if possible.

Answers

1. 10, 17
2. 4, 9
3. 4, 13; 3, 3; 21
4. 2, 5

5. 2

6. 6, 0, 0, 4, 2, 4

7. 10

8. 1610, 1912

9. 1824

10. 66.26 mi, no

Solutions

1. Let b and a be the numbers of amendments in the Bill of Rights and amendments added to the Constitution after the Bill of Rights. Then $\begin{cases} a + b = 27 \\ a^3 + b^3 = 5913 \end{cases}$. Let us consider the second equation and use the formula of the sum of cubes (see Appendix I-AF) to obtain $a^3 + b^3 = (a + b)(a^2 - ab + b^2) = 27(a^2 - ab + b^2) = 5913 \Rightarrow a^2 - a$ $(27 - a) + (27 - a)^2 = 219 \Rightarrow a^2 - 27a + 170 = 0$. The last quadratic equation has two solutions $a = 10$ and $a = 17$. The corresponding $b = 17$ and $b = 10$. We take $b = 10$ and $a = 17$, because $b < a$.

The Bill of Rights consists of 10 amendments. 17 amendments have been added to the constitution after the Bill of Rights.

2. Let us present $\frac{1}{n \cdot (n+1) \cdot (n+2)}$ as $\frac{A}{n} + \frac{B}{n+1} + \frac{C}{n+2} = \frac{A(n+1)(n+2) + Bn(n+2) + Cn(n+1)}{n \cdot (n+1) \cdot (n+2)}$ from which follows that $A = 1/2$, $B = -1$, $C = 1/2$. Thus $\frac{16}{n \cdot (n+1) \cdot (n+2)} = 8 \cdot \left(\frac{1}{n} - \frac{2}{n+1} + \frac{1}{n+2} \right)$ and

$\frac{16}{1 \cdot 2 \cdot 3} + \frac{16}{2 \cdot 3 \cdot 4} + \frac{16}{3 \cdot 4 \cdot 5} + \frac{16}{4 \cdot 5 \cdot 6} + \frac{16}{5 \cdot 6 \cdot 7} + ... + \frac{16}{n \cdot (n+1) \cdot (n+2)} + ... =$

$8 \cdot \left(\frac{1}{1} - \frac{2}{2} + \frac{1}{3} + \frac{1}{2} - \frac{2}{3} + \frac{1}{4} + \frac{1}{3} - \frac{2}{4} + \frac{1}{5} + \frac{1}{4} - \frac{2}{5} + \frac{1}{6} + \frac{1}{5} - \frac{2}{6} + \frac{1}{7} + ... \right)$

that after cancellation leads to 4.

The eighth term is $\frac{16}{8 \cdot 9 \cdot 10} = \frac{1}{9 \cdot 5}$. One-fifth of its reciprocal is 9.

There are four time zones in the continental United States and nine time zones in the United States and its territories.

3. The variables x, y, and z are nonnegative integers. Let us perform division:

$\frac{172}{53} = 3 + \frac{13}{53} = 3 + \frac{1}{\frac{53}{13}} = 3 + \frac{1}{4 + \frac{1}{13}} = x + \frac{1}{y + \frac{1}{z}}$. Then $x = 3$, $y = 4$, $z = 13$.

Three states have the longest commonly used names that contain 13 letters: Massachusetts, North Carolina, South Carolina.

Iowa, Ohio, and Utah have the shortest state names with 4 letters. The longest official state name with 21 letters belongs to the State of Rhode Island.

4. Let $x > 0$ be the size of the smallest side of a right triangle. Since all sides of the triangle are consecutive terms of an arithmetic series (see Appendix I-SS), the other sides are $x + d$ and $x + 2d$, where $d > 0$ is the difference of an arithmetic series. The triangle is right, then, by the Pythagorean theorem, $x^2 + (x + d)^2 = (x + 2d)^2 \Rightarrow x = 3d$ or $d = -x$. The last option $d = -x$ does not satisfy $x > 0$, $d > 0$. Thus, the sides of the triangle are x, $\frac{4x}{3}$, $\frac{5x}{3}$.

The center of the circumcircle is at the middle of the hypotenuse, then its radius is $R = \frac{5x}{6}$. Applying properties of a tangent line (a tangent line to a circle is perpendicular to the radius drawn to the point of tangency, and two tangents from the same point to a circle are equal (see Appendix I-CP)), we can write $\frac{5x}{3} - (x - r) = \frac{4x}{3} - r \Rightarrow r = \frac{x}{3}$, where r is the radius of the inscribed circle. Then, $\frac{r}{R} = \frac{2}{5}$, and the numbers of states can be 2 and 5, 4 and 10, 6 and 15, and so on. The product of the first pair is the smallest.

There are other ways to find x, for instance, combining different formulas for the area of a right triangle. Indeed, from $A = pr$ and $A = bh/2$ we can get $r = \frac{x}{3}$, where b and h are two legs of a right triangle.

The chickadee is a state bird of two states (Maine and Massachusetts). The mockingbird is a state bird of five states (Arkansas, Florida, Mississippi, Tennessee, and Texas).

5. From the quadratic equation $x^2 - 5x + p = 0$ follows $\begin{cases} x_1 x_2 = p \\ x_1 + x_2 = 5 \end{cases}$.

Squaring both sides of the second equation $5^2 = (x_1 + x_2)^2 = x_1^2 + 2x_1 x_2 + x_2^2$ and using the second equation, we obtain $x_1^2 + x_2^2 = 5^2 - 2p$. Then, from $x_1^3 + x_2^3 = (x_1 + x_2)(x_1^2 - x_1 x_2 + x_2^2) = 5(5^2 - 3p) = 95$ follows $p = 2$ (see Appendix I-AF).

Two states: Mississippi has chosen magnolia as its state flower and state tree. Virginia has adopted American dogwood as its state flower and state tree.

6. Rearranging terms in $xy - x - 3y = 0$ we obtain $xy - x - 3y + 3 = 3 \Rightarrow (x - 3)(y - 1) = 3$. Because x and y are integers, then the only possible options are

$$\begin{cases} x - 3 = 1 \\ y - 1 = 3 \end{cases} \Rightarrow \begin{matrix} x = 4 \\ y = 4 \end{matrix}; \begin{cases} x - 3 = 3 \\ y - 1 = 1 \end{cases} \Rightarrow \begin{matrix} x = 6 \\ y = 2 \end{matrix}; \begin{cases} x - 3 = -1 \\ y - 1 = -3 \end{cases}$$

$$\Rightarrow \begin{matrix} x = 2 \\ y = -2 \end{matrix}; \begin{cases} x - 3 = -3 \\ y - 1 = -1 \end{cases} \Rightarrow \begin{matrix} x = 0 \\ y = 0 \end{matrix}.$$

Because x and y should be nonnegative, then the pair (2, –2) does not work. The maximum value of x is 6, of y is 4. Their minimum values are 0 and 0 and medians are 4 and 2.

No states have chosen either the sequoia or mountain ash tree as their state trees. Two states have chosen magnolia as its state flower (Louisiana, Mississippi). Four states have a rose (Georgia-Cherokee rose, Iowa-white rose, New York-rose, North Dakota-wild prairie rose). Four states have the word "blossom" in its state flower (Arkansas, Michigan-apple blossom, Delaware-peach blossom, Florida-orange blossom). Six states have chosen an oak (Iowa-oak, Connecticut, Illinois, Maryland-white oak, Georgia-live oak, New Jersey-red oak).

7. Let $\log_a(\sqrt{x^2 + 1} + x) = f(x)$. Let us check whether the function $f(x)$ is an odd or even function (see Appendix I-AF):

$f(-x) = \log_a(\sqrt{(-x)^2 + 1} + (-x)) = \log_a(\sqrt{x^2 + 1} - x) = \log_a \frac{(\sqrt{x^2+1} - x)(\sqrt{x^2+1} + x)}{\sqrt{x^2+1} + x}$
$= \log_a \frac{1}{\sqrt{x^2+1} + x} = \log_a(\sqrt{x^2 + 1} + x)^{-1} = -\log_a(\sqrt{x^2 + 1} + x) = -f(x)$. Hence, f (x) is an odd function and from $f(-x) = \log_a(\sqrt{x^4 + 1} - x) = -10x$ follows $f(x) = \log_a(\sqrt{x^4 + 1} + x) = 10$.

Ten states are: Alabama (southern pine), Arkansas (pine tree), Idaho (western white pine), Maine and Michigan (white pine), Minnesota (Norway pine), Montana (Ponderosa pine), Nevada (Bristlecone pine), New Mexico (pinon pine), North Carolina (longleaf pine).

8. Let the first two digits of the year when Santa Fe became a capital a, the last two digits of the year when New Mexico became a state b, and c be two legs and hypotenuse of a right triangle, $a > b$, then

$$\begin{cases} \frac{ab}{2} = 96 \\ a + b + c = 48 \\ a^2 + b^2 = c^2 \end{cases} \Rightarrow \begin{cases} 2ab = 384 \\ (a + b)^2 = (48 - c)^2. \\ a^2 + b^2 = c^2 \end{cases}$$

Adding the first and third equations leads to $a^2 + b^2 + 2ab = c^2 + 384$, that, in combination with the second equation, produces $c^2 + 384 = (48 - c)^2$. The last equation has a solution of $c = 20$. Then $a + b = 28$ and $ab = 192$ give pairs of solutions $(a, b) = \{(16, 12), (12, 16)\}$. Since $a > b$, we take $a = 16$, $b = 12$.

Santa Fe became the capital of Mexico in 1610. Mexico became a state in 1912.

9. Let n and d be the first term and difference of an arithmetic sequence, then $n^2 + (n + d)^2 = (n + 2d)^2 \Rightarrow 3d^2 + 2nd - n^2 = 0 \Rightarrow d = -n$, $n/3$. The solution $d = -n$ does not satisfy the problem and $d = n/3$ produce integers if n is divisible by 3. Let us consider different options of Pythagorean triples:

n	d		sum	n	d		sum
3	1	(3, 4, 5)	12	12	4	(12, 16, 20)	48
6	2	(6, 8, 10)	24	15	5	(15, 20, 25)	60
9	3	(9, 12, 15)	36	18	6	(18, 24, 30)	72

The smallest sum with 7 among its digits occurs at $n = 18$ and the middle term is $18 + 18/3 = 24$.

Tallahassee became the capital of the Florida Territory in 1824.

10. Let a, b, and c be the distances between Boston and other three state capitals, then $\begin{cases} \sqrt[3]{abc} = 63 \\ \frac{1}{a} + \frac{1}{b} + \frac{1}{c} = 0.05 \\ a^2 + b^2 + c^2 = 14509 \end{cases} \Rightarrow \begin{cases} abc = 63^3 \\ \frac{ab + ac + bc}{abc} = 0.05 \\ a^2 + b^2 + c^2 = 14509 \end{cases}$.

Substituting the results from the system to $(a + b + c)^2 = a^2 + b^2 + c^2 + 2(ab + ac + bc) \Rightarrow (a + b + c)^2 = 14509 + 2 \cdot 0.05 \cdot 63^3 \Rightarrow \frac{a + b + c}{3} = \frac{198.78}{3} = 66.26$.

The average distance between Boston and Harford, Providence, and Concord, is 66.26 miles. The provided information is not sufficient to determine the distance between Boston to other three state capitals.

II

Geography of the United States

II.1 Geography of US States

The United States of America is the largest country in the Western Hemisphere and the fourth largest country in the world. Its 50 states have different elevations, areas, and divisions.

1. All the US states are divided into smaller administrative units. Such subdivisions are called *counties* in the 48 US states, *parishes* in Louisiana, and *boroughs* in Alaska. A *county equivalent* is referred to parish, borough, independent city, and census area.

   As of 2013, there were 313 more counties than 6/7 of the total number of counties and county-equivalents, 5 more boroughs and 3 less census areas than 2/449 of the total number of counties and county-equivalents, 8 more parishes and 15 less independent cities than 8/449 of the total number of counties and county-equivalents, and one District of Columbia. How many counties and county-equivalents were in the USA in 2013? How many parishes, boroughs, census areas, and independent cities did the USA have in 2013?

2. The number of US states with the most border states is the smallest number in the domain of the function $y = \sqrt[4]{3 - |x - 5|}$, while the largest value in the domain is the number of states they are bordering with. How many states have the most border states? How many states do they have a border with? Name these states.

3. The number of US states with the lowest elevation above 1000 ft is the value of p, such that the quadratic equation $x^2 + px + 4 = 0$ has two integer solutions. Moreover, the equation will have integer solutions if an integer is added to both coefficients 4 and p. How many states have the lowest elevation above 1000 ft? Name these states if any.

4.–6. *Elevations of US states differ significantly. The highest elevation in Florida is the lowest highest elevation among all the US states. Louisiana has the same*

DOI: 10.1201/9781003229889-2

39

average elevation as Florida, while the average elevation in Delaware is lower. However, the highest elevations of Delaware and Louisiana are greater than in Florida. The lowest elevations of Louisiana and California are lower than the lowest elevations in Florida. Nevertheless, Florida is the flattest state in the USA with the lowest range of elevations.

4. The highest elevation of Florida is presented by a three-digit number. These digits form a set of Pythagorean triples written in the increasing order. The average elevation of Florida is also presented as a three-digit number. Its first digit is the radius of the circle inscribed in the triangle with sides equaled the digits of the highest elevation. Each following digit of the average elevation is one less than the preceding digit or 0 if the difference is negative. The lowest elevation is the mean of the last two digits of the average elevation. What are the highest, lowest, and average elevations in Florida?

5. The State of Delaware has the *lowest average elevation* among all the US states. It can be found as a solution to the equation

$$\frac{(2+3)x}{3^2} \log_{\sqrt[6]{(x-33)^5}} (2^2 \cdot 3x + 3^2)^{\frac{3}{x}} = 2^2.$$

What is the average elevation of the State of Delaware?

6. The State of Colorado has the *highest average elevation* among all the US states. Its average elevation in feet is $(x^2 + y^2)(\sqrt{x^2} + \sqrt{y^2})^2$, where x and y satisfy the equation $x^2 - y^2 = 60$, and y is a positive integer less than 14. What is the average elevation of State of Colorado?

7. The Territory Northwest of the River Ohio or the *Northwest Territory* became a territory of the USA on July 13, 1787, after being a part of the British Province of Quebec and a territory under British rule. The southeastern portion of the territory was admitted to the Union as the state of Ohio on March 1, 1803. The Northwest Territory covered the current states of Ohio, Indiana, Illinois, Michigan, Wisconsin, and the northeastern part of Minnesota.

The average area of six states that came from Northwest Territory taken from one source is 63,195.7 sq. mi and the median area is 60,162 sq. mi. When all areas of states are verified at another source, it is found that only the areas of Indiana and Ohio coincide with the areas given on the first source. The areas of Illinois, Minnesota, and Wisconsin are 3089, 7326, and 58 sq. mi larger, while the area of Michigan is 1274 sq. mi smaller. Based on this information, is it possible to calculate the average and median of the data from the second source?

8. The areas of four states Virginia, Washington, Georgia, and Texas (listed from the smallest to the largest area) rounded to the nearest thousands square miles are prime numbers. They are composed of at least two digits from the following set of nine digits {1, 2, 3, 4, 5, 6, 7, 9, 9}. The sum of three digits of the area value of one state is 17. All sums of digits of the values of other three areas are greater than 6. The last digits of two-digit areas form consequent terms of a geometric sequence. Find the areas of Virginia, Washington, Georgia, and Texas as well as the area of North Dakota, which is approximately the same as the area of Washington.

9. The areas (rounded to the nearest thousands square miles) of five states Pennsylvania, Georgia, Utah, Michigan, Nevada (from smallest to the largest area) form an arithmetic sequence. Moreover, if it were a state with the area of 72 thousand sq. mi, then the areas of this state (put between two areas) and the five mentioned states would form six consequent terms of that arithmetic sequence. The total area (the sum of areas of six states) would become 471 thousand sq. mi. Find the areas of Pennsylvania, Georgia, Utah, Michigan, Nevada as well as areas of Oregon and Wyoming which are approximately equal to the area of Michigan. Round all area values to the nearest thousands.

10. Colorado and Wyoming are two neighboring states with a rectangular shape. Wyoming is 80 mi longer than Colorado. The width of Wyoming is the same as the length of Colorado. The area of Colorado is more than the area of Wyoming by 5600 sq. mi. Total border length of both states (sum of their perimeters) is 2600 mi. What are the lengths and the widths of Colorado and Wyoming?

11. In addition to its mainland, Louisiana has many islands that cover nearly 1.3 million acres.

   How many islands does Louisiana have? Use the following hints to answer the question.

   A. The number of islands is a four-digit number
   B. At most two digits of the number of islands are the same
   C. The second digit is not the same as the last one
   D. The sum of all even digits is 16
   E. The product of the fourth (ones) digit and a two-digit number, formed by the second (hundreds) digit as tens and the first digit as ones, is the same as another two-digit number composed of the third digit as tens and the second digit as ones

F. The first three digits of the number of islands are consequent terms an increasing geometric sequence

12. The area of water resources in Alaska is the *largest* in the USA, while water resources take most of the total area in Michigan. The total area of Michigan is 1971 sq. mi larger than the area of water resources in Alaska. The land area in Alaska is 5251 sq. mi more than 10 times the land area in Michigan. The difference between the land and water areas in Michigan is 16,364 sq. mi. The sum of land areas in both states is 627,180 sq. mi. What percentage of the total area is taken by water resources in Alaska and Michigan?

13. The *driest* US state is New Mexico with 0.2% of the total area of water resources. Can we conclude that New Mexico has the lowest area of water resources among all the US states?

Answers

1. 3143, 3007, 19, 11, 64, 41
2. 2 states with 8 bordering states
3. 5
4. 345 ft, 0 ft, 100 ft
5. 60 ft
6. 6800 ft
7. yes, 64,728.0 sq. mi, no
8. 43 thousand sq. mi, 59 thousand sq. mi, 71 thousand sq. mi, 269 thousand sq. mi
9. 46 thousand sq. mi, 59 thousand sq. mi, 85 thousand sq. mi, 98 thousand sq. mi, 111 thousand sq. mi
10. 380 mi by 280 mi, 280 mi by 360 mi
11. 2482
12. 14.2, 42.5
13. no

Solutions

1. Let x be the total number of counties and county-equivalents, then $\frac{6}{7}x + 313$ is the number of counties, $\frac{2}{449}x + 5$ is the number of boroughs, $\frac{2}{449}x - 3$ is the number of census areas, $\frac{8}{449}x + 8$ is the

number of parishes, $\frac{8}{449}x - 15$ is the number of independent cities, which, with 1 for the District of Columbia, lead to

$$x = 1 + \left(\frac{6}{7}x + 313\right) + \left(\frac{2}{449}x + 5\right) + \left(\frac{2}{449}x - 3\right) + \left(\frac{8}{449}x + 8\right)$$
$$+ \left(\frac{8}{449}x - 15\right),$$

$$\Rightarrow x = \frac{2834}{3143}x + 309 \Rightarrow \frac{309}{3143}x = 309 \Rightarrow x = 3143.$$

There are 3007 counties, 19 boroughs, 11 census areas, 64 parishes, 41 independent cities, and the District of Columbia among 3143 counties and county-equivalents.

2. The domain of $y = \sqrt[4]{3 - |x - 5|}$ is $|x - 5| \le 3 \Rightarrow -3 \le x - 5 \le 3 \Rightarrow 2 \le x \le 8$ (see Appendix I-AV). Thus, the lowest number is 2 and the largest is 8.

Two states, Missouri and Tennessee, are bordering with 8 states. Missouri has borders with Arkansas, Illinois, Iowa, Kansas, Kentucky, Nebraska, Oklahoma, Tennessee. Tennessee borders with Alabama, Arkansas, Georgia, Kentucky, Mississippi, Missouri, North Carolina, Virginia.

3. Integer solutions (if any) to the equation $x^2 + px + 4 = 0$ may be from the set $\{\pm 1, \pm 2, \pm 4\}$. Following the statement of the problem, the number of states p should be nonnegative, $p \ge 0$. Then the signs of coefficients are +++, and by Descartes rule of sign, there are no positive roots (see Appendix I-PE). The integer negative roots can be among $\{-1, -2, -4\}$. By Vieta's theorem (see Appendix I-QE), the roots x_1 and x_2 of the equation $x^2 + px + 4 = 0$ satisfy $x_1 + x_2 = -p$ and $x_1 \cdot x_2 = 4$. As required in the problem, for any integer n there should be integer solutions x_{1n} and x_{2n} to the equation $x^2 + (p + n)x + (4 + n) = 0$, or $x_{1n} + x_{2n} = -(p + n)$ and $x_1 \cdot x_2 = 4 + n$.

Let us consider different options:

1. $x_1 = -1$, then $x_2 = -4$ (from $x_1 \cdot x_2 = 4$), $p = 5$ (from $x_1 + x_2 = -p$), and $x_{1n} = -1$, $x_{2n} = -(4 + n)$ are integers for any natural n.

2. $x_1 = -2$, then $x_2 = -2$ and $p = 4$ and the equation $x^2 + (p + n)x + (4 + n) = 0$ has $D = (4 + n)^2 - 4(4 + n) = n(4 + n)$, which is not a square for any n. Thus, this option does not work.

3. The solution $x_1 = -4$ leads to the same results as in (1), i.e., $x_1 = -1$ because then $x_2 = -1$ and $p = 5$.

The lowest elevations of five states, Colorado, Montana, New Mexico, Utah, and Wyoming, are above 1000 ft.

4. The only Pythagorean triple with one-digit numbers (3, 4, 5) leads to the highest elevation 345 ft. The triangle with sides 3, 4, and 5 is a right triangle.

1st way: The area of the triangle is $A = 6$ sq. ft, its half perimeter is $p = 6$ ft, then from $A = pr$ (see Appendix I-T) follows $r = 1$.

2nd way: From the properties of tangent lines to a circle (see Appendix I-CP): $4 - r = 5 - (3 - r) \Rightarrow r = 1$.

Thus, the first digit of the average elevation is 1, the second digit is $0 = 1 - 1$, the third digit is also 0 because $0 - 1 < 0$. The average between the last two digits is 0.

The lowest elevation in Florida is 0 ft, its average elevation is 100 ft, and its highest elevation is 345 ft.

5. The domain of the equation $\frac{(2+3)x}{3^2} \log_{\sqrt[6]{(x-33)^5}}(2^2 \cdot 3x + 3^2)^{\frac{3}{x}} = 2^2$ is $x > 33$, $x \neq 34$ (see Appendix I-EL). Using properties of logarithms, the equation can be simplified to $\frac{5x}{9} \frac{3}{x} \frac{6}{5} \log_{(x-33)}(12x + 9) = 4$, and then to $\log_{(x-33)}(12x + 9) = 2$. Applying a relation between exponential and logarithmic functions (see Appendix I-EL), we can rewrite the last expression as $(x - 33)^2 = (12x + 9)$, which is a quadratic equation $x^2 - 78x + 1080 = 0$. Its solution is $x = 18$ and $x = 60$. The first number 18 is not in the domain.

The State of Delaware has an average elevation of 60 ft.

Remark. It is interesting to note that 60 ft is 18 m, which appeared in the problem as another possible solution to the quadratic equation.

6. From the equation $x^2 - y^2 = 60$ and the statement that y is a positive integer we conclude that x is also an integer (positive or negative) and both x and y are odd or even at the same time. Then $(x + y)$ and $(x - y)$ are even. Since $x^2 - y^2 = (x - y)(x + y) = 60$, then 60 has to be presented as a product of two even integer factors. The options are: (2, 30), (6, 10), (-2, -30), (-6, -10). Because y is a positive integer less than 14, then neither (2, 30) nor (-2, -30) works. Two pairs (6, 10) and (-6, -10), are left, i.e., $\begin{cases} x - y = 10 \\ x + y = 6 \end{cases}$ or $\begin{cases} x + y = 10 \\ x - y = 6 \end{cases}$. The solution to the first system (8, -2) does not work.

The solution to the second one is (8, 2). Similarly, (–6, –10) lead to (–8, 2). Both (8, 2) and (–8, 2) produce the same result: $(x^2 + y^2)(\sqrt{x^2} + \sqrt{y^2})^2 = (x^2 + y^2)(|x| + |y|)^2 = (8^2 + 2^2)(8 + 2)^2 = 6800$.

The average elevation of the highest US state, State of Colorado, is 6800 ft (2100 m).

7. The difference is $3089 - 1274 + 7326 + 58 = 9199$, then $9199/6 = 1533.167$ added to 63,195.7 gives the new average area of 64,728.8. Based on the provided information it is not possible to find the median.

State	Area Difference
Illinois	3089
Indiana	None
Michigan	–1274
Minnesota	7326
Ohio	None
Wisconsin	58
Mean	64,728.9

8. If areas of all states are at least two-digit numbers, then the area of only one state, which is Texas, is a three-digit number with 17 as the sum of its digits. The sum of two sets of numbers, 1, 7, 9 and 2, 6, 9, is 17.

If 1, 7, 9 form the area of Texas, then the rest of the digits, 2, 3, 4, 5, 6, 9 are left for two-digit numbers for other three areas that should be prime, which is impossible. Indeed, 2, 4, 5, and 6 cannot be the second (ones) digit, because then the corresponding number will be divisible by at least by 2 or 5. Then only 3 and 9 are left for the units digit of a two-digit prime number, but three are needed.

Thus, the set {2, 6, 9} is left. It produces just one prime 269.

From the remaining digits 1, 3, 4, 5, 7, 9 for the areas of three states, only 1, 3, 9 form a geometric sequence. Then 4, 5, and 7 are left for the tens digit of each area.

If 4 is taken as the first digit, then it can form 41, 43, and 49, from which 41 does not work (the sum of the digits is less than 6) and 49 is not prime. Thus, 43 is left.

If 5 is taken, then it can form 51 and 59, from which 51 is not prime. Hence, 59 is left and 71 is the area of the third state.

The problem leads to the same result if 5 or 7 is considered first.

The area of Virginia is 43 thousand sq. mi (actual 42,774 sq. mi), the area of Georgia is 59 thousand sq. mi (59,425 sq. mi), the area of Washington and North Dakota is 71 sq. mi (actual 71,362 and 70,762, respectively), and the area of Texas is 269 thousand sq. mi (268,820 sq. mi).

9. Let x thousand sq. mi be the smallest area and d be the difference of the arithmetic sequence. As stated in the problem, x and d are positive integers. Then the sum of six consequent terms of the sequence is $x + (x + d) + (x + 2d) + (x + 3d) + (x + 4d) + (x + 5d) = 471$.

Let us solve this equation for d: $2x + 5d = 157 \Rightarrow d = \frac{157 - 2x}{5} \Rightarrow d = 31 + \frac{2(1 - x)}{5}$. The term $(1 - x)$ should be divisible by 5 or $x = 5k + 1$. The corresponding $d = 31 - 2k$, $k = 1, 2, \ldots, 15$. Next, the number 72 is one of six terms of the arithmetic sequence, then $72 = x + nd \Rightarrow 71 = 5k + 31n - 2kn$, $1 \le n \le 6$.

Let us consider different values of n, $1 \le n \le 6$:
If $n = 1$, then $40 = 3k$ and k is not integer.
If $n = 2$, then $k = 9$.
If n takes values 3, 4, 5, 6, then we will get $22 = 5k$, $53 = 3k$, $84 = 5k$, or $115 = 7k$ correspondingly. None of them produce an integer k.

Therefore, $n = 2$ and $k = 9$ and 72 is the third term of the arithmetic sequence. The first term is 46 and the difference of the arithmetic sequence is 13. The terms are 46 (Pennsylvania), 59 (Georgia), 72, 85 (Ohio), 98 (Michigan, Oregon, and Wyoming), 111 (Nevada).

The area of Pennsylvania is 46 thousand sq. mi (46,055 sq. mi), the area of Georgia is 59 thousand sq. mi (59,425 sq. mi), the area of Utah is 85 thousand sq. mi (84,899 sq. mi), the areas of Michigan (97,990 sq. mi), Oregon (98,466 sq. mi), and Wyoming (97,818 sq. mi) are approximately 98 thousand sq. mi, the area of Nevada is 111 thousand sq. mi (110,567 sq. mi).

10. Let x and y be the width and length of Wyoming and u and v be the width and length of Colorado, then

$$\begin{cases} uv - xy = 5600 \\ 2(u + v) + 2(x + y) = 2600 \\ v = x \\ y - v = 80 \end{cases}.$$

Substitution of $x = v$ and $y = 80 + v$ from the third and fourth equations to the first and second equations reduces the original nonlinear system of four equations in four variables to a system of two equations in two variables:
$\begin{cases} uv - v(v + 80) = 5600 \\ u + v + (v + v + 80) = 1300 \end{cases}$ or

$$\begin{cases} (1220 - 3v)v - v(v + 80) = 5600 \\ u = 1220 - 3v \end{cases}$$, that can be reduced further to the

quadratic equation $v^2 - 285v + 1400 = 0$. The solution of the quadratic equation is $v = 280$ and $v = 5$. The solution $v = 5$ does not satisfy the applied meaning of the problem. Then $u = 380$, $x = 280$, $y = 360$.

Colorado is 380 mi by 280 mi, Wyoming is 280 mi by 360 mi.

11. Let a, b, and c be the first three digits that represent the number of islands (Hint A), $b = aq$, $c = aq^2$ (Hint F), where q is the ratio of a geometric sequence that cannot be 1 (Hint B). This gives us the following options for the first three digits: 124, 139, and 248 with the sum of even digits of 6, 0, and 14 leaving 10, 16, and 2 for the last digits (Hint D). Thus, the four-digit number is 2482.

Louisiana has 2482 islands.

The problem can be solved if other hints are used.

12. Let A, L, and W be the total area, land area, the water area of Alaska (with subscript A) and Michigan (with subscript M), then

$$\begin{cases} A_M - W_A = 1971 \\ L_A - 10L_M = 5251 \\ L_M - W_M = 16364 \\ L_A + L_M = 627180 \end{cases} \Rightarrow \begin{cases} A_M - W_A = 1971 \\ A_A - W_A - 10A_M + 10W_M = 5251 \\ A_M - W_M - W_M = 16364 \\ A_A - W_A + A_M - W_M = 627180 \end{cases}$$

$$\underset{\frac{1}{11}(Eq.II-Eq.IV)+Eq.III}{\Rightarrow} \begin{cases} W_M = 40175 \\ (Eq.\ III) \Rightarrow A_M = 96,714 \\ (Eq.\ I) \Rightarrow W_A = 94,743 \\ (Eq.\ I) \Rightarrow A_A = 665,384 \end{cases}$$

Hence, $W_M/A_M \approx 0.4154$, $W_A/A_A \approx 0.1424$.

13. No, the total area is also important. For instance, New Mexico's 0.2% of the total area for the water resources is 292 sq. mi, while West Virginia's 0.8% is 192 sq. mi and District of Columbia's 10.3% is 7 sq. mi of water resources, which are less than New Mexico's 292 sq. mi.

II.2 Highest Peaks and Lowest Elevations

The US is famous for its gorgeous landscapes and unique natural formations. Mountain peaks and lowest elevation points are gems of national parks and national monuments. *Many mountain peaks and giant rocks are sacred for native people and appear in their legends. The weather is also quite diverse across different parts of the USA.*

1. *Mount McKinley* or *Denali,* meaning "The Great One" in Athabascan, is the highest mountain peak in the USA and North America. Its peak is a centerpiece of Denali National Park and Preserve in Alaska.

   The first and fourth (tens) digits as well as the second (thousands) and fifth (ones) digits of the five-digit height of Mount McKinley in feet are the same. The tens as x and hundreds as y are prime numbers that satisfy $y^4 - 4x^4 = 17$. The second digit is one less than the difference between the third and fourth digits. How tall is Mount McKinley?

2. *Mount McKinley,* Alaska, is the highest summit in the USA and in the North America, while *Mount Whitney,* California, is the highest point in the 48 US contiguous states.

   The arithmetic mean of elevations of Mount McKinley and Mount Whitney is 17,412.5 ft. Their geometric mean is 17,168.04 ft. How tall are these mounts? Round the result to the nearest integers.

3. *Guadalupe Peak* or *Signal Peak* belongs to the Guadalupe Mountains located in New Mexico and Texas. The peak is the highest natural point in Texas and a jewel of Guadalupe Mountains National Park.

   The elevation of Guadalupe Peak in feet is a four-digit number. The differences between two consecutive digits of the number starting with the highest (thousands) position are consecutive powers of the same base. The mean of all digits is not their median, which is not a prime but an integer. What is the elevation of Guadalupe Peak?

4. Colorado has the *most* fourteeners or mountains with peaks above 14,000 ft in the contiguous US states.

   The number of Colorado fourteeners is the difference between the first and the last terms of the expansion of $(3\sqrt[3]{2} + 1)^n$. The third term of this expansion is related to its fourth term as $3^2 \cdot \sqrt[3]{2}$ to 1. How many fourteeners are in Colorado?

5. The topological elevation of a summit is the vertical distance above sea level, while the topological prominence is its elevation relative to the surrounding terrain.

The number of mountain peaks in Colorado with at least 500 m (1640.4 ft) of topological prominence is an integer term of the expansion of $\left(\sqrt{11} + \frac{1}{\sqrt[4]{44}}\right)^5$ into a sum. Each of these summits exceeds 4000 m (13,123.4 ft) of topologic elevation. How many mountain peaks in Colorado are above 500 m of topological prominence?

6. Mountain peaks of the island of Hawaii are sacred for Hawaiians. An ancient law allows only high-ranking tribal chiefs to visit the peak of *Mauna Kea*, or *White Mountain* in Hawaiian. Its dormant volcano is the highest point in Hawaii. Mauna Kea has been a National Natural Landmark since 1972.

   Mauna Kea could be the tallest mountain on Earth if measured from its oceanic base. The mountain is 5500 ft more below sea level than above it. What is the elevation of Mauna Kea above sea level and from its ocean base if the numbers in feet below (as y) and above (as x) sea level lie on the line that passes through the point (1000, 100) perpendicular to $3y + 2x = 5$?

7. *Devils Tower* in Crook County, Wyoming, rises like a giant above the surrounding land. Being sacred to native tribes, this natural monolith is called *Bear Lodge* or *Brown Buffalo Horn* by Lakota people and *Bear's House, Bear's Lair, Home of bears* by Crow tribe. An 1875 expedition misinterpreted *Bad God's Tower* as *Devil's Tower*. Indeed, the rock sparks red as the breath of the Devil.

   There are many legends on how the rock was formed. In one legend, two girls were chased by giant bears and Great Spirit saved them. In another, two boys spotted by bears were saved by the Creator. In all legends the rock rose from the ground to the sky and bears tried to climb from every side leaving huge scratch marks. These parallel cracks made Devils Tower one of the finest crack climbing in North America.

   Devils Tower was the first to be declared as the United States Natural Monument by President Theodore Roosevelt in 1906. Devils Tower appeared in *Surreal Places You Need to Visit Before You Die*, movie *Close Encounters of the Third Kind*, and other media.

   The summit of Devils Tower is 3847 ft higher above sea level than above the surrounding terrain. Its height above sea level is 46 ft more than four times its height above the surrounding terrain. How tall is Devils Tower above the surrounding terrain and sea level?

8. *Chimney Rock*, or *Chimney Tower*, or *Elk's Peak* is a rock formation in western Nebraska. It was first mentioned in 1827 as a remarkable

pointer for travelers along the Oregon, California, and Mormon Trails. Chimney Rock became a Historic National Site on August 9, 1956.

Its peak is $\overline{abcd}$ feet above the sea level. The sum of two middle digits of its elevation $\overline{abcd}$ is its first digit and the sum of the first two digits is the last digit. The elevation $\overline{abcd}$ is an even number with a nonzero product of all digits. Its second digit is not less than the third one.

Moreover, the Chimney Rock is $\overline{\left(\frac{ab}{2}\right)(c+d)\left(\frac{b-c}{2}\right)}$ feet above the North Platte River Valley and its splendor spire rises $\overline{\left(\frac{a+b}{2}\right)\left(\frac{b+c}{2}\right)\left(\frac{a+d}{2}\right)}$ feet from its conical base.

How tall are the spire and Chimney Rock above sea level and the river valley?

Remark: $\overline{abcd}$ denotes a four-digit number to tell it apart from the product $abcd$.

9. *Mount Elbert* is the highest summit of the Rocky Mountain in North America, the second-highest mountain in the contiguous USA, and the highest of the fourteeners of Colorado. It is located in the San Isabel National Forest. The mountain was named after the territorial governor Samuel Hitt Elbert of Colorado in 1873.

   An active volcano *Mount Rainier* in Mount Rainier National Park, Washington, is the most glaciated peak in the contiguous USA. *Mount Whitney* in California is the highest summit in the contiguous USA and the Sierra Nevada. The west slope of the mountain is in Sequoia National Park and the east slope is in the Inyo National Forest. Surprisingly, its summit is very close to the lowest point of the North America.
   These three peaks are the highest points in their states.

   The elevation of Mount Elbert above sea level is 722/725 of the elevation of Mount Whitney. It is 10 ft lower than the average elevation of the three peaks, Mount Elbert, Mount Whitney, and Mount Rainier. Mount Elbert is 30 ft higher than Mount Rainier. What is the elevation of each mountain peak above sea level?

10. The distance in miles between the highest summit in the contiguous USA, Mount Whitney (14,510 ft) in California, and its lowest elevation (270 ft below sea level), which is in Death Valley National Park, California, can be presented by a number with tens digit taken as an integer cubed and its units as that integer

squared. A decimal of the number is the mean between its tens and units digits. None of the digits in the distance value are repeated. What is the distance between these two extreme points?

11. Mount Whitney 4421 m, California, is the highest summit in the Sierra Nevada and the contiguous USA. Surprisingly, it is just 136.2 km of the lowest point (85 m below sea level) of North America, which is in Death Valley National Park. If a straight line could connect the summit with the lowest point, then approximately how far from the lowest point will be the point of intersection of this line with the horizontal line between these extreme points taken at sea level?

12. The hottest temperature in the USA was recorded in Greenland Ranch in the Death Valley National Park, California on July 10, 1913. The coldest temperature was recorded at Prospect Creek, Alaska, on January 23, 1971. These bring the range of possible temperatures in the USA to 214 F, while the mean of the record hottest and coldest temperatures is 27 F. What are the hottest and coldest temperatures recorded in the USA?

13. Only two US states, California and Louisiana, have their lowest elevation points below sea level. The lowest point in California is in Badwater Basin, Death Valley National park. This point is also the lowest point in the Western Hemisphere. The lowest point of Louisiana is in New Orleans.

    None of three digits that represent the California lowest elevation in feet is 1. The second digit of its lower elevation is twice the first digit raised to the third (ones) digit power. The second digit of the value of the lower elevation in California also represents the value below sea level of the lowest point in Louisiana. What are the lowest elevations in California and Louisiana?

14. The largest hailstone in the nation fell in Coffeyville, Kansas. It was 44.5 cm in the circumference of the largest cross section. The density of solid ice is 0.919 g/cm^3. Find the volume, surface area, and mass of this hailstone. Assume that the hailstone had a perfect spherical shape.

Answers

1. 20,320 ft
2. 20,320 ft, 14,505 ft
3. 8751 ft
4. 53

5. 55

6. 13,800 ft, 33,100 ft

7. 1267 ft, 5114 ft

8. 325 ft, 4226 ft, 480 ft

9. 14,500 ft, 14,440 ft, 14,410 ft

10. 84.6 mi

11. 2.6 km

12. −80F, 134F

13. −282 ft, −8 ft

14. 1488 cm^3, 630 cm^2, 1397 kg   .

Solutions

1. Factoring the equation $y^4 - 4x^4 = 17$, we get $(y^2 - 2x^2)(y^2 + 2x^2) =$
   17·1, that leads to $\begin{cases} y^2 - 2x^2 = 1 \\ y^2 + 2x^2 = 17 \end{cases}$ or $\begin{cases} y^2 - 2x^2 = 17 \\ y^2 + 2x^2 = 1 \end{cases}$. Factors −17 and
   −1 are not considered because $y^2 + 2x^2$ cannot be negative. The
   second system does not have an integer solution, the first system
   gives $\begin{cases} y^2 = 9 \\ x^2 = 4 \end{cases}$ that leads to 3 as the third digit and 2 as the first and
   forth digits. Then, $3 - 2 - 1 = 0$ gives the second and the fifth digits.

   Mount McKinley, Alaska, is 20,320 ft tall.

2. Let u and v be the elevations of Mount McKinley and Mount
   Whitney, then $\begin{cases} \frac{u+v}{2} = 17, 412.5 \\ \sqrt{uv} = 17, 168.04 \end{cases} \Rightarrow \begin{cases} u + v = 34, 825 \\ uv = 294, 741, 597.4 \end{cases}$!. Then,
   u and v are solutions to $x^2 - 34, 825x + 294, 741, 597.4 = 0$ (see
   Appendix I-QE), presented by $x = \frac{34,825 \pm 5815}{2}$ or $x = \{20,320, 14,505\}$.

   Mount McKinley is 20,320 ft and Mount Whitney is 14,505 ft.

3. Let a, b, c, d are the digits from of the elevation $\overline{abcd}$ of the
   Guadalupe Peak in feet and x be the base, then
   $\begin{cases} a - b = x^y \\ b - c = x^{y+1} \\ c - d = x^{y+2} \end{cases} \Rightarrow a - d = x^y + x^{y+1} + x^{y+2} = x^y(1 + x + x^2)$, which should
   be a one-digit number. Let us consider different options:

   If $x = 0$, then for any $y > 0$, the elevation $\overline{aaaa}$ has the same mean
   and median of the digits.

If $x = 1$, then for any y, the elevation $\overline{a(a-1)(a-2)(a-3)}$ has noninteger median of all digits.

If $x = 2$, then $a - d = 2^y \cdot 7$ and y can be only 0. Thus, the system leads to the following numbers 7640, 8751, or 9862. The mean and median of digits of 7640 and 9862 are not the same. Only 8751 has 6 as the median of its digits, which is not a prime. x cannot be equal to or greater than 3 to form a four-digit number because then $x^y(1 + x + x^2) = 3^y(1 + 3 + 3^2)$ is not a one-digit numbers even for $y = 0$.

Thus, the only option is 8751.

The elevation of the Guadalupe Peak is 8751 ft.

4. Using the Newton's binomial expansion (see Appendix I-AF) and remembering that counting starts from 0, the third and fourth terms of $(3\sqrt[3]{2} + 1)^n$ are T_2 and T_3 and $\frac{T_2}{T_3} = \frac{\frac{n!}{2!(n-2)!}(3\sqrt[3]{2})^{n-2}}{\frac{n!}{3!(n-3)!}(3\sqrt[3]{2})^{n-3}} = \frac{3(3\sqrt[3]{2})}{(n-2)} = \frac{3^2 \cdot \sqrt[3]{2}}{1}$.

Therefore $n = 3$ and $(3\sqrt[3]{2})^3 - 1^3 = 53$.

There are 53 fourteeners in Colorado.

5. 1st way: From Pascal's triangle or Newton's binomial expansion (see Appendix I-AF): $\left(\sqrt{11} + \frac{1}{\sqrt[4]{44}}\right)^5 = 121\sqrt{11} + 5 \cdot 121\frac{1}{\sqrt[4]{44}} + 10 \cdot 11\sqrt{11}\frac{1}{2\sqrt{11}} + 10 \cdot 11 \cdot \frac{1}{\sqrt[4]{44^3}} + 5 \cdot \sqrt{11}\frac{1}{44} + \frac{1}{44\sqrt[4]{44}}$, has an integer term $10 \cdot 11\sqrt{11}\frac{1}{2\sqrt{11}} = 55$.

2nd way: According to Newton's binomial expansion, each term of $\left(\sqrt{11} + \frac{1}{\sqrt[4]{44}}\right)^5$ is presented as $\binom{5}{k}(\sqrt{11})^{5-k}\left(\frac{1}{\sqrt[4]{44}}\right)^k = \binom{5}{k}\frac{11^{\frac{10-3k}{4}}}{2^{\frac{k}{2}}}$, $k = 0, 1, 2, 3, 4, 5$. Both $\frac{10-3k}{4}$ and $\frac{k}{2}$ should be integers, which is possible only for $k = 2$. Then $\binom{5}{2}(\sqrt{11})^{5-2}\left(\frac{1}{\sqrt[4]{44}}\right)^2 = 55$.

55 Colorado mountain peaks are at least 500 m of topological prominence.

6. The equation of the line that passes through (1000, 100) perpendicular to $3y + 2x = 5$ is $y = 3/2x - 1400$, then $\begin{cases} y - x = 5500 \\ -2y + 3x = 2800 \end{cases}$. Multiplying the first equation by 2 and adding it to the second equation gives $x = 13,800$. Substituting this value to the first equation leads to $y = 19,300$.

Mauna Kea is 13,800 ft are above sea level and its base-to-peak height is 33,100 ft.

7. Let h and H feet be the height of Devils Tower above the surrounding terrain and sea level, then

$$\begin{cases} H - h = 3847 \\ 4h + 46 = H \end{cases}$$. Substitution of H from the second equation to the first one leads to $3h = 3801$, from which $h = 1267$ and $H = 5114$.

Devils Tower is 3801 ft above the surrounding terrain and 5114 ft above sea level.

8. $\begin{cases} b + c = a \\ a + b = d \end{cases} \Rightarrow 2b + c = d$. The number is even then d is even, and c is even. The product of all digits is nonzero $\Rightarrow b \neq 0, c \neq 0, d \geq 4$, i.e., d can be 4, 6, or 8. Then possible options of $\overline{abcd}$ with the second digit not less than the third one are 4226 and 5328. Only 4226 leads to positive digits in

$$\left(\frac{ab}{2}\right)(c + d)\left(\frac{b - c}{2}\right) = 480 \text{ and } \left(\frac{a + b}{2}\right)\left(\frac{b + c}{2}\right)\left(\frac{a + d}{2}\right) = 325$$

The Chimney Rock is 4226 ft above sea level and 480 ft above the river valley. Its spire rises 325 ft from its conical base.

9. Let e, r, and w be the elevation of Mount Elbert, Mount Rainier, and Mount Whitney, then $\begin{cases} e = 722/725w \\ e - 30 = r \\ e = (r + e + w)/3 - 10 \end{cases} \Rightarrow$

$$\begin{cases} e = 722/725w \\ e - 30 = r \\ 3e = e - 30 + e + 725/722e - 30 \end{cases} \Rightarrow \begin{cases} e = 14,440 \\ r = 14,410 \\ w = 14,500 \end{cases}.$$

The elevations of Mount Elbert is 14,440 ft, of Mount Rainier is 14,410 ft, and of Mount Whitney is 14,500 ft.

10. 1 and 2 are the only integers that have one-digit numbers as their cubes. If 1 is taken, then the resulting number 11.1 has repeated digits. If 2 is chosen, then it leads to 84.6 mi.

The highest summit in the contiguous USA is just 84.6 mi from the lowest elevation.

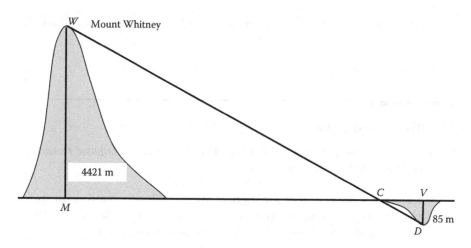

FIGURE II.1

11. Triangles $\triangle WMC$ and $\triangle DVC$ are similar (see Figure II.1). Let x be the distance from the crossing point C and the lowest point D, then:

$$\frac{85}{4421} = \frac{x}{136,\,200 - x} \Rightarrow 4506x = 11,\,577,\,000 \Rightarrow x = 2569\text{m}.$$

Only 2569 m are between the lowest and highest US points.

12. Let h and c be the hottest and coldest temperatures
$$\Rightarrow \begin{cases} h - c = 214 \\ \frac{h+c}{2} = 27 \end{cases} \Rightarrow \begin{array}{l} h = 134 \\ c = -80 \end{array}.$$
The temperature in the USA ranges from –80 F to 134 F.

13. The only option for the second digit to satisfy all conditions of the problem is $8 = 2 \cdot 2^2$.

The lowest point in California is 282 ft deep, and in Louisiana is 8 ft.

14. $C = 2\pi r \Rightarrow r = \frac{C}{2\pi} \Rightarrow V = \frac{4}{3}\pi r^3 = \frac{4}{3}$
$$\pi \left(\frac{44.5}{2\pi}\right)^3 \approx 1488 \text{ cm}^3; \; S = 4\pi r^2 = 4\pi \left(\frac{44.5}{2\pi}\right)^2 \approx 630 \text{ cm}^2.$$

$M = 0.919 \cdot V = 1367.5$ g.

The largest hailstone had a volume of 1488 cm^3, surface area of 630 cm^2, and weighted 1367.5 g.

II.3 Rivers and Lakes

Rivers, lakes, and springs are precious national water resources, valuable recreation facilities, and jewels of any region.

1. Kingsley Lake is one of the highest lakes in Florida famous for its clear water, swimming, skiing, and fishing. Air pilots call it the Silver Dollar Lake because of its sparkled circular form.

   Let us assume that infinitely many equilateral triangles are inscribed in the circle formed by the Kingsley Lake. Describe the common region of these triangles. Find its area if the difference between the areas of the lake and an inscribed equilateral triangle is $\frac{25}{16}\left(\pi - \frac{3\sqrt{3}}{4}\right)$ sq. mi? What is the diameter of Kingsley Lake?

2. The Grand Prismatic Spring in the Midway Geyser Basin is one of many treasures of the first US national park, Yellowstone National Park, located Wyoming, Montana, and Idaho. The spring is the largest hot spring in the USA and the third largest hot spring in the world after Frying Pan Lake in New Zealand and Boiling Lake in Dominica. Pigmented bacteria grow around edges producing the unique range of vivid colors, while extreme heat makes the center of the pool sterile. The spring is featured in *Surreal Places You Need to Visit Before You Die* and many other media programs.

   European explorers discovered the Grand Prismatic Spring in 1839, noticing a "boiling lake" with a diameter of 300 ft. It was later recorded that the spring is approximately 250 by 300 ft. Assuming a circular shape in the first case and a rectangular shape in the second case, which of the following is correct?

   A. The area and circumference of a circular spring are less than the area and perimeter of its rectangular shape.

   B. The area of a circular spring is less than the area of its rectangular shape but the circumference of a circular spring is greater than the perimeter of its rectangular shape.

   C. The area of a circular spring is greater than the area of its rectangular shape, but the circumference of a circular spring is less than the perimeter of its rectangular shape.

D. The area and circumference of a circular spring are greater than the area and perimeter of its rectangular shape.

3. The Hamilton pool in Texas is a scenic natural swimming pool surrounded by a gorgeous grotto, huge limestones, and large stalactites. It was created thousands of years ago when the dome of an underground river collapsed. The beautiful waterfall from the Hamilton Creek down to the pool never dries up keeping its water level constant. The Hamilton pool belongs to the Balcones Canyonlands Preserve. The pool is also listed in many sources including *Surreal Places in America You Need to Visit before You Die*.

How tall is the waterfall above the Hamilton pool if its height equals $M \cdot \overline{MY}$ feet, where $\overline{MY}$ is a two-digit number with digits M and Y defined by $\overline{MY} = Y^M$?

4. Native Americans have lived on the central Oregon cost for more than 6000 years. To preserve the historic significance and highlight the unique beauty of the area, the US Forest Service developed the 2700-acre Cape Perpetua Scenic Area. The cape was named by Captain James Cook who discovered it on St. Perpetua's Day on March 7, 1778.

The Cape Perpetua coastline is famous for its mysterious places such as the Devil's Churn, Spouting Horn at Cook's Chasm, and Thor's Well. A unique feature of the area is a 600-year-old *Giant Sitka Spruce* known as the *Silent Sentinel of the Siuslaw* and designated as a *"Heritage Tree"*.

The *Heritage Tree* stands 185 ft high and has a 40-ft circumference at its base. Assuming a conic shape of the tree, the volume of its trunk is

A. 15,711 cubic ft

B. 2467 cubic ft

C. 7400 cubic ft

D. 7852 cubic ft

E. 98,667 cubic ft

5.–6. *Lake Superior is the largest of the Great Lakes, the largest freshwater lake in the world by surface area, the world third-largest freshwater lake by volume, and the largest lake in North America by volume. The lake is shared by Ontario, Canada, and Minnesota, Wisconsin, and Michigan. The area of the* *lake is much larger than the area of the biggest island in the USA, the island of Hawaii or the "Big Island".*

5. The areas of Lake Superior and Big Island of Hawaii in thousand square miles are the largest and smallest possible values that a periodic function with the period $T = 2\frac{1}{2}$ can take at $x = 2$ if the function satisfies the equation

$$2f^2\left(-\frac{31}{2}\right) - 24f\left(-\frac{21}{2}\right) + 62 = f^2\,(12) + 12f\,(22) - 66.$$

What are the areas of the island of Hawaii and Lake Superior in thousand square miles?

6. The third, first, and fourth (ones) digits of the four-digit area of the Big Island of Hawaii are the first three consequent terms of a geometric sequence. Its second, third, and first digits form the first three consequent terms an arithmetic sequence and the last (ones) digit is its fifth term. The ratio of the geometric sequence and the difference of the arithmetic sequence are the same. Find the area of the Big Island of Hawaii.

7. The total area of New York City, NY, is 468.4 sq. mi. Its land area is 29 sq. mi less than twice its water area. As estimated in 2012, 8,336,697 people live in New York City making it the most populous city in the United States. What percentage of the city area is the land area and water area? What is the population density in New York City?

8. The lengths of $a(b + c) + c$ rivers in the USA are at least 500 mi long and a of them cross or form international borders. The Yukon and Columbia rivers are among b rivers that begin in Canada and flow into the USA while the Milk, Saint Lawrence, and Red River are among c rivers that flow from the USA into Canada. The Colorado and Rio Grande rivers are among b rivers that flow from the USA into Mexico or form a border between these two countries. The Milk River is among $(c - b)$ rivers that cross the international border twice. Find the number of US rivers which are at least 500 mi, the number of rivers that cross the border between the USA and Canada or Mexico at least once or twice if a, b, and c are coefficients of equations of the curves

$$ax + by = 3, \; y = cx^2 + bx - a, \; x^4 + y^3 + a = 0$$

that intersect at the point $(1, -2)$.

9. The Missouri and Mississippi Rivers are the longest main-stem rivers in the USA. The Yukon River that flows through Alaska, Yukon Territory, and British Columbia, Platte River in Colorado, Wyoming, and Nebraska, and Tanana River in Alaska are also among 30 longest rivers in the USA.

Yukon River is twice as long as Platte River and three times as long as Tanana River. Platte is 330 mi longer than Tanana. How long are Yukon River, Platte River, and Tanana River?

10. The Colombia River is the largest river in the Pacific Northwest region of North America and the fourth-largest river in the USA by volume. It has the greatest flow among North American rivers entering the Pacific. The Colombia River is one of two rivers that flows from Canada into the USA. The Saint Lawrence River is one of three rivers that begins in the USA and flows into Canada. It is the longest Atlantic Ocean stem river.

The Colombia River is 43 mi longer than twice the Saint Lawrence River. Five times the length of the Saint Lawrence River is 514 mi more than twice the length of the Colombia River. How long are the Colombia and Saint Lawrence Rivers?

11. An impact crater is a circular depression in a solid body formed by a smaller body coming with velocity over 10,000 ft/s. The Weaubleau crater in western Missouri is one of the 50 known impact craters on the Earth and the fourth largest one in the USA. Moreover, it is the largest exposed impact crater in the nation because the other three have been buried or are under water. The crater was named after nearby Weaubleau Creek.

If a scalene triangle inscribes in the circular Weaubleau crater then the areas of three triangles obtained by connecting the center of the circle with the vertices of the triangle will be 51, 75, and 78 sq. ft. Find the diameter of the Weaubleau crater.

Answers

1. A circle, $\frac{25}{64}\pi$ sq. mi, 2.5 mi
2. A
3. 50 ft
4. D
5. 4 thousand sq. mi, 32 thousand sq. mi

6. 4028 sq. mi, yes

7. 64.6%, 35.4%, 27,550

8. 38, 7, 1

9. 1980 mi, 990 mi, 660 mi

10. 1243 mi, 600 mi

11. 12 ft

Solutions

1. Another circle which is inscribed in any inscribed equilateral triangle will be a common part of all equilateral triangles. From properties of inscribed circles, its radius r is a half of the radius of the circumscribed circle R, which is Kingsley Lake. The height of a triangle and the radius of the lake are connected as (see Appendix I-CP, T) $R = \frac{2}{3}h; \quad h = \frac{a\sqrt{3}}{2}$.

   Then $\quad A_o - A_\Delta = \pi R^2 - \frac{ah}{2} = R^2\left(\pi - \frac{3\sqrt{3}}{4}\right) = \frac{25}{16}\left(\pi - \frac{3\sqrt{3}}{4}\right) \Rightarrow R = \frac{5}{4}$.

   The area of Kingsley Lake $A = \pi R^2 = \frac{25}{16}\pi$. The center of the inner circle coincides with the center of Kingsley Lake and its radius $r = 5/8$ mi. The area of Kingsley Lake $A = \frac{25}{16}\pi$ sq. mi and its diameter is 2.5 mi.

2. The area and circumference of a circular spring with the diameter 300 ft are (see Appendix I-CP) $C = 2\pi r \approx 942$ft; $A = \pi r^2 \approx 70,686$sq. ft. The area and perimeter of the rectangular 250 by 300 shape are $P = 2(250 + 300) = 1100$ft and $A = 250{\cdot}300 = 75,000$sq. ft.The area and circumference of a circular spring is less than the area and perimeter of its rectangular shape.

3. Let us consider different options for Y or M. It is easy to see that neither Y nor M are 0 or 1. If $M = 2$, then Y can be 5 or 6, but only 5 satisfies $5^2 = 25$. Other values of M do not lead to a solution.

   Another way to solve the puzzle is to locate a square number between 20 and 29 for $M = 2$, which is 25, that gives the value to MY. If $M = 3$ or larger, then there is no number with the third or higher power represented as a two-digit number and such that $MY = Y^M$.

4. The volume of a cone $V = \frac{1}{3}\pi r^2 h$, where the radius of the base is found from $C = 2\pi r = 40$ as $r = \frac{20}{\pi}$. Hence, $V = \frac{20^2\pi \cdot 185}{3\pi^2} \approx 7852$ cubic ft.

5. Because the function $f(x)$ is a periodic function with the period of $T = 2\frac{1}{2} = 5/2$, then $f\left(-\frac{31}{2}\right) = f\left(2 - 7{\cdot}\frac{5}{2}\right) = f(2); f\left(-\frac{21}{2}\right) =$

$f\left(2 - 5\cdot\frac{5}{2}\right) = f(2); f(12) = f\left(2 + 4\cdot\frac{5}{2}\right) = f(2); f(22) = f\left(2 + 8\cdot\frac{5}{2}\right) = f(2).$

From $2f^2\left(-\frac{31}{2}\right) - 24f\left(-\frac{21}{2}\right) + 62 = f^2(12) + 12f(22) - 66$ follows $2f^2(2) - 24f(2) + 62 = f^2(2) + 12f(2) - 66 \Rightarrow f^2(2) - 36f(2) + 128 = 0 \Rightarrow f(2) = 4; f(2) = 32.$

Thus, the lowest value 4 is the area in thousand square miles of the Big Island of Hawaii and 32 is the area of the Lake Superior in thousand square miles.

Remark. The function $f(x)$ at $x = 2$ can take either 4 or 32, otherwise it will not be a function (it will not pass the vertical line test).

6. Let a be the third digit of the area and d be the common difference of the arithmetic and ratio of the geometric sequences. Then the second, first, and fourth digits of the area are the first, third, and fifth terms of an arithmetic sequence, i.e., $a - d$, $a + d$, and $a + 3d$. The first and fourth digits of the area are the second and third terms of a geometric sequence, i.e., ad and ad^2. Then $a + d = ad$, and $a + 3d = ad^2$ and combining them, we obtain $\frac{3d}{d} = \frac{a(d^2 - 1)}{a(d - 1)}$. Hence, $d + 1 = 3$ or $d = 2$. Thus, $a = 2$ and the four-digit number is 4028.

The area of the Big Island of Hawaii is 4028 thousand sq. mi.

7. Let l and w be the land and water areas, then $l + 29 = 2w$ and $l + w = 468.4$. Subtracting the first equation from the first one and dividing both sides of the obtained equation by 3, we get $w = 165.8$ and $l = 302.6$. Hence, $302.6/468.4 = 64.6$, and $8,336,697/302.6 = 27,550$.

The land takes 64.6% and the water takes 35.4% of the total area. The population density in New York City is 27,550 people per sq. mile.

8. Substituting $(1, -2)$ to the equations $ax + by = 3$, $y = cx^2 + bx - a$, $x^4 + y^3 + a = 0$, we obtain $a - 2b = 3, -2 = c + b - a$, and $1 - 8 + a = 0$, which lead to $a = 7$, $b = 2$, $c = 3$.

The lengths of 38 rivers in the USA are at least 500 mi long and 7 of them cross international borders. Only seven rivers cross the international border at least once. The Yukon, Columbia Milk, Saint Lawrence, and Red River cross the border between the USA and Canada, while the Colorado and the Rio Grande rivers form a border with Mexico. The Milk River is the only river that crosses the international border twice.

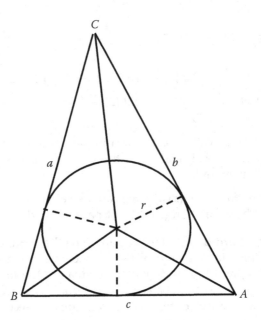

FIGURE II.2

9. Let x be the length of the Yukon River, then, remembering that the Yukon River is twice as long as the Platte River and three times as the Tanana River, we get $\frac{x}{2} - \frac{x}{3} = 330 \Rightarrow x = 1980$.

The Yukon River is 1980 mi-long. The Platte River is 990 mi and the Tanana river is 660 mi.

10. Let x be the length of the Saint Lawrence River and y be the length of the Colombia River. From $2x + 43 = y$ and $5x - 514 = 2y$ follow $x = 600$ and $y = 1243$.

The Colombia River is 1243 mi-long (2000 km) and the Saint Lawrence River is 600 mi (970 km).

11. The inradius r in the triangle ABC (see Figure II.2) is perpendicular to each side (see Appendix I-CP, T). Therefore, it is a height in the constructed triangles. The areas of the three triangles are

$$A_a = \frac{ar}{2} = 51, \ A_b = \frac{br}{2} = 75, \ A_c = \frac{cr}{2} = 78 \Rightarrow a = \frac{2 \cdot 51}{r}, \ b = \frac{2 \cdot 75}{r}, \ c = \frac{2 \cdot 78}{r},$$

where a, b, and c are sides of the triangle, and A_a, A_b, and A_c are the areas of the three triangles with the corresponding side as the base. The area of the inscribed triangle is $A = A_a + A_b + A_c = 204$ sq. ft and its half perimeter is $p = \frac{204}{r}$ ft. The Heron's Formula for the area leads to

$$204 = \sqrt{\frac{204}{r}\left(\frac{204}{r} - \frac{2 \cdot 51}{r}\right)\left(\frac{204}{r} - \frac{2 \cdot 75}{r}\right)\left(\frac{204}{r} - \frac{2 \cdot 78}{r}\right)} = \frac{7344}{r^2} \Rightarrow r = 6\text{ft}.$$

The Weaubleau crater is 12 ft in diameter.

II.4 Mysterious Places

It is believed that paranormal activities occur in the USA and all over the world. People, cattle, ships, and planes have mysteriously disappeared in these regions. UFOs, "bigfoot", giant snakes, and other strange phenomena have been observed there. The supernatural spots vary in their shape and size. The spot in California's Redwood forest is circular, while the Bermuda, Bridgewater, Massachusetts, and Alaska Triangles have a triangular form. There is much mystery behind such phenomena, though researchers state that paranormal activities are fabricated and inaccurately reported. However, all these extraterrestrial phenomena and anomalous zones are very popular discussion topics.

1. *The Bermuda Triangle* is a triangular area on the Atlantic Ocean with imaginary vertices in Bermuda, Florida, and Puerto Rico. It is thought that dozens of ships and planes have mysteriously vanished without any trace. The Bermuda Triangle legend started in 1950s with reports of Edward Van Winkle Jones on disappearances of ships and planes. This area is also referred as the *Devil's Triangle* because some believe that Devil plays there.

   The Bermuda triangle is defined differently in various sources. Taken its dimensions as in Figure II.3, find the ratio and product of the circumradius and inradius. Can the problem be solved without finding both radii?

2.–3. *The Michigan Triangle is a mysterious water triangle over the center of Lake Michigan. Strange objects and phantom planes have been seen in the area since disappearance of Captain George during a routine sail in 1937. Thirteen years later, Northwest Airlines Flight 2501 from New York City to Minneapolis vanished while passing the Michigan Triangle. Although some debris had been located, the rest of the wreckage and bodies of passengers have not been found yet.*

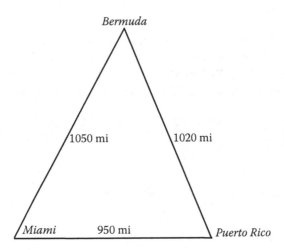

FIGURE II.3

2. Ludington MI, Benton Harbor MI, and Manitowoc WI are the vertices of the Michigan Triangle. Its sides are 61, 127, and 149 mi. Describe the triangle according to its sides and internal angles.

3. Ludington, MI; Benton Harbor, MI; Manitowoc, WI are three corners of the Michigan Triangle. These cities are separated by Lake Michigan shown in Figure II.4.

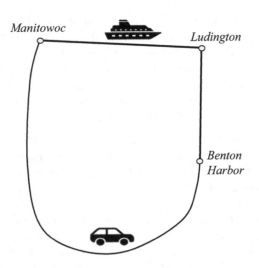

FIGURE II.4

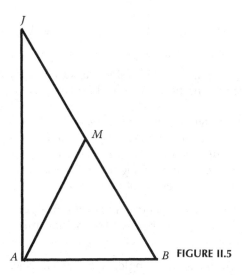

B **FIGURE II.5**

Andrew and Bill from Manitowoc were invited by their friend Phillip to visit him in Ludington. Andrew and Bill wanted to arrive in Ludington at the same time. However, Andrew preferred driving, while Bill decided to take a ferry that left Manitowoc at 2 pm CT. The ferry arrived in Ludington at 7 pm ET. Andrew left Manitowoc 3 hours before Bill and drove with the planned speed until he reached Benton Harbor. His drive between Benton Harbor and Ludington was 10 mph slower than planned. Therefore, Andrew arrived 30 minutes after Bill. The driving distance between Manitowoc and Benton Harbor is 30 mi less than twice the driving distance between Benton Harbor and Ludington. What are the driving and straight distances (taking the ferry) between Manitowoc and Ludington if the driving distance is 7 times the straight distance?

4.–5. *An unusually high number of people have vanished in the Alaska's Bermuda Triangle that includes a vast area of unexplored wilderness, sprawling forests, craggy mountain peaks, and barren tundra. Some stories are included in the documentary Alaska's Bermuda Triangle.*

4. The perimeter of the Alaska's Triangle is 2525 mi. Its vertices A, J, and B are in Anchorage (the most populous city of Alaska), capital Juneau, and Barrow (the largest city of the North Slope Borough), shown in Figure II.5. The sides AJ and BA are related to as 23 to 34, and BA to JB as 17 to 22. Describe the triangle according to its sides and internal angles. Find the sides and angles of the Alaska's triangle. Is all information given in the problem needed to answer these questions?

5. The vertices A, J, and B of the Alaska's triangle are in Anchorage, Juneau, and Barrow. The sides AJ and BA are related as 23 to 34,

and BA to JB as 17 to 22. The median AM from the angle A to the side BJ is $25\sqrt{\frac{717}{2}}$ mi. Describe the triangle according to its sides and internal angles. Find the sides and angles of the Alaska's triangle. Is there any extra (unused) information given in the problem?

6.–9. *The mysterious Bridgewater Triangle in southeastern Massachusetts is believed to be a site of UFOs, balls of fire, sightings of poltergeists, bigfoot, mutilation of cattle, and other supernatural phenomena. In his book "Mysterious America", paranormal researcher Loren Coleman defined boundaries of the Bridgewater Triangle by the towns of Freetown, Abington, and Rehoboth as the vertices of the triangle.*

> Note: The straight distances between endpoints of the Bridgewater Triangle are taken in the problems below. The actual area and distances vary in difference sources.

6. The perimeter of the Bridgewater Triangle, which is isosceles, is 60 mi. One side is twice the second one. What are the sides of the Bridgewater Triangle? Is the answer unique?

7. The area of the Bridgewater Triangle is $36\sqrt{15}$ sq. mi. The triangle is isosceles with one side twice the second one. What are the sides of the triangle?

8. It takes 5 minutes less than 2 hours to make the 74-mile Bridgewater triangle loop, Abington–Freetown–Rehoboth–Abington (see Figure II.6). The drive from Rehoboth to Abington takes 5 minutes less than from Abington to Freetown and 10 minutes more than from Freetown to Rehoboth. The driving distances from Abington to

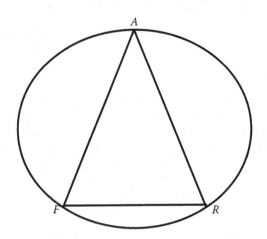

FIGURE II.6

Freetown, from Freetown to Rehoboth, and from Rehoboth to Abington are related to as 15 to 8 to 14. What is the average speed on each segment of the road?

9. The straight distances between Abington and Freetown and between Abington and Rehoboth are the same and twice the distance between Freetown and Rehoboth, $AF = AR = 2FR$ (see Figure II.6). Which of the following statements about $\triangle AFR$ are correct? Justify your choice.

A. The length of the arc AF is twice the arc FR

B. The angle AFR is twice the angle FAR, $\angle AFR = 2\angle FAR$

C. $\sin \angle AFR = 2\sin \angle FAR$

D. $\cos \angle AFR = 2\cos \angle FAR$

E. The incenter and the circumcenter coincide

F. The median, bisector, and height from the angle $\angle FAR$ coincide

G. The median, bisector, and height from the angle $\angle AFR$ coincide

10. A mystery spot in the heart of California's Redwood Forest was discovered in 1939. According to its owners, it is a *gravitational anomaly* where the laws of physics and gravity do not work. Strange phenomena and various illusions are demonstrated there. The values of the area and circumference of the circular mystery California spot are related to as 75 to 2. Find the diameter of the mystery California spot.

A. 37.5

B. 75

C. 75

D. 150

E. not possible

Answers

1. 2, 166,345, yes

2. Scalene and acute triangle

3. 60 mi, 420 mi

4. Scalene and obtuse triangle, no, 575 mi, 850 mi, 1100 mi, yes

5. Scalene and obtuse triangle, no, 575 mi, 850 mi, 1100 mi, yes

6. 12 mi, 24 mi, 24 mi, yes

7. 12 mi, 24 mi, 24 mi

8. 40 mph, 32 mph, 42 mph

9. C, F

10. D

Solutions

1. Let us consider three formulas for the area of a triangle (see Appendix I-T) $A = \frac{a+b+c}{2}r$, $A = \frac{abc}{4R}$, and $A = \sqrt{\frac{a+b+c}{2} \cdot \frac{-a+b+c}{2} \cdot \frac{a-b+c}{2} \cdot \frac{a+b-c}{2}}$, where the a, b, c are sides of the triangle, and r and R are its inradius and circumradius. From $A \cdot A = A^2$, the ratio between the inradius and circumradius is found from $\frac{a+b+c}{2} \cdot \frac{-a+b+c}{2} \cdot \frac{a-b+c}{2} \cdot \frac{a+b-c}{2} = \frac{(a+b+c)r}{2} \cdot \frac{abc}{4R}$

$\Rightarrow \dfrac{R}{r} = \dfrac{abc}{4} \cdot \dfrac{2}{-a+b+c} \cdot \dfrac{2}{a-b+c} \cdot \dfrac{2}{a+b-c} \Rightarrow \dfrac{R}{r} = \dfrac{2 \cdot 950 \cdot 1030 \cdot 1020}{960 \cdot 940 \cdot 1100} \Rightarrow \dfrac{R}{r}$

$= \dfrac{1,996,140,000}{992,640,000} \approx 2.$

The product of inradius and circumradius is

$\dfrac{a+b+c}{2}r = \dfrac{abc}{4R} \Rightarrow rR = \dfrac{abc}{2(a+b+c)} \Rightarrow rR = \dfrac{950 \cdot 1030 \cdot 1020}{2 \cdot 3000} = 166,345.$

It is not important to know the values of inradius and circumradius to solve the problem.

2. The length of sides of the triangle are not the same, then it is a scalene triangle.

   Considering the internal angles: If c is the length of the longest side of a triangle and $a^2 + b^2 > c^2$, where a and b are the lengths of the other sides, then the triangle is an acute triangle (see Appendix I-T). In our case: $61^2 + 127^2 > 149^2$.

   The Michigan Triangle is a scalene acute triangle.

3. Since 2 pm CT is 3 pm ET, the ferry trip was 4 hours. Let s and v be the distance between Benton Harbor to Ludington and the planned average driving speed, then the distance between Manitowoc and Benton Harbor is $2s - 30$, and $\begin{cases} \dfrac{s}{v-10} - \dfrac{s}{v} = \dfrac{1}{2} \\ \dfrac{s+2s-30}{v} = 4+3 \end{cases}$. Substituting $s = \frac{1}{20}v(v-10)$ from the first equation to the second we obtain the quadratic equation $3v^2 - 170v - 600 = 0$. Putting everything

together, we obtain the system $\begin{cases} s = \frac{1}{20}v(v - 10) \\ 3v^2 - 170v - 600 = 0 \end{cases}$, from which

follows $v = 60$, $s = 150$ (see Appendix I-QE).

Thus, the total driving distance is 420 mi, and the straight distance is $420/7 = 60$ mi. The second solution of the quadratic equation $v = -20/6$ does not satisfy the applied meaning of the problem.

The driving distance between Manitowoc and Ludington is 420 mi. The straight distance is 60 mi.

4. Since $AJ/BA = 23/34$ and $BA/JB = 17/22 = 34/44$, then $AJ = 23s$, $BA = 34s$, $JB = 44s$, where s is any natural number. The lengths of sides of the triangle are not the same, then it is a scalene triangle.

   The perimeter of the triangle $2525 = 23s + 34s + 44s \Rightarrow s = 25$. Hence, the sides of the triangle are 575, 850, and 1100 miles. Since the internal angles $(23s)^2 + (34s)^2 < (44s)^2$, one interior angle of the triangle is greater than 90°, and it is an obtuse triangle.

   Angles of a triangle can be found by the law of cosine (see Appendix I-T), $c^2 = a^2 + b^2 - 2ab \cos\angle C$ as $\angle B = 31.06°$, $\angle J = 49.70°$, $\angle A = 99.24°$.

   The value of the perimeter is not needed to describe the triangle, but it is vital to find numerical values of sides and angels.

5. Since $AJ/BA = 23/34$ and $BA/JB = 17/22 = 34/44$, the lengths of sides of the triangle are not the same. Hence, it is a scalene triangle.

   Combining $AJ = 23s$, $BA = 34s$, $JB = 44s$ with the formula for the median of a triangle (see Appendix I-QE) $4m_{BJ}^2 = 2BA^2 + 2AJ^2 - BJ^2$, we get $4 \cdot 25^2 \cdot \frac{717}{2} = 2 \cdot 23^2 s^2 + 2 \cdot 34^2 s^2 - 44^2 s^2 \Rightarrow s = 25$. Hence, the sides of the triangle are 575, 850, and 1100 mi.

   By the internal angles: If JB is the length of the longest side, then $AJ^2 + BA^2 < JB^2$, or $(23s)^2 + (34s)^2 < (44s)^2$. Then one interior angle of the triangle is greater than 90°. Hence, the triangle is an obtuse triangle. Angles of a triangle can be found by the law of cosine, $c^2 = a^2 + b^2 - 2ab \cos \angle C$ as $\angle B = 31.06°$, $\angle J = 49.70°$, $\angle A = 99.24°$.

   The value of the perimeter is not needed to describe the triangle, though it is vital to find numerical values of sides and angels.

6. Let a be one side of the triangle. It is impossible to have a triangle with a as another side and $2a$ as the base because then the sum of two sides of a triangle is not greater than the third side, $a + a = 2a$. Hence, a is the base, $2a$ is a side, $a + 2a + 2a = 60$, and $a = 12$ is the only choice.

7. The sum of two sides of a triangle should be greater than the third side. Thus, the only choice is that two sides of the isosceles triangle are twice its base a. Then the height of the triangle:

$$h = \sqrt{(2a)^2 - \left(\frac{a}{2}\right)^2} = \frac{\sqrt{15}\,a}{2} \text{ and } A = \frac{a^2\sqrt{15}}{4} = 36\sqrt{15} \Rightarrow a = 12.$$

The sides of the Bridgewater triangle (straight distances) are 12 mi, 24 mi, and 24 mi.

8. The driving distances from Abington to Freetown, from Freetown to Rehoboth and from Rehoboth to Abington are related to as 15/8/14. The loop is 74 mi long, then $15s + 8s + 14s = 74 \Rightarrow s = 2$. Thus, the distances between Abington and Freetown, Freetown and Rehoboth, and Rehoboth and Abington are 30 mi, 16 mi, and 28 mi.

Next, 5 minutes is 1/12 of an hour, 10 minutes is 1/6 of an hour, and 5 minutes less than 2 hours is $115/60 = 23/12$ hours.
Let t hours be the driving time from Rehoboth to Abington, then $\left(t + \frac{1}{12}\right) + \left(t - \frac{1}{6}\right) + t = \frac{23}{12}$. Driving from Rehoboth to Abington takes $t = \frac{2}{3}$ hours, from Abington to Freetown $t + \frac{1}{12} = \frac{3}{4}$ hours, and Freetown to Rehoboth $t - \frac{1}{6} = \frac{1}{2}$ hours. Therefore, the average speed from Abington to Freetown is $\frac{30}{3/4} = 40$ mph, from Freetown to Rehoboth is $\frac{16}{1/2} = 32$ mph, and from Rehoboth to Abington is $\frac{28}{2/3} = 42$ mph.

The driving distances between Abington and Freetown, Freetown and Rehoboth, and Rehoboth and Abington are 30 mi, 16 mi, and 28 mi.

9. The median, bisector, and height from the vertex in an isosceles triangle coincide, then F is correct.

According to the law of sines (see Appendix I-T), $\frac{\sin \angle AFR}{AR} = \frac{\sin \angle FAR}{FR} \Rightarrow \frac{\sin \angle AFR}{2FR} = \frac{\sin \angle FAR}{FR}$, which leads to $\sin \angle AFR = 2\sin \angle FAR$, then C is correct.

The centers of inscribed circle and the circumscribed circle coincide only in equilateral triangles.

10. Following formulas for the area and circumference of a circle (see Appendix I-CP), we get $\frac{A}{C} = \frac{\pi r^2}{2\pi r} = \frac{r}{2} = \frac{75}{2} \Rightarrow r = 75 \Rightarrow d = 150$. The answer is D.

III

National and State Parks

III.1 National Parks

Each nation proudly develops and protects national parks and monuments. All of them are carefully chosen for their unique natural beauty, historical importance, and scientific significance.

1. The first national park in the United States of America, Yellowstone National Park, was signed into law by President Ulysses S. Grant in 1872. This national park is located in Wyoming, Montana, and Idaho.

   How many parks (the total number) were open in 1872, if this number is $\frac{x-y}{2}$, where the integers x and y satisfy $(x^2 - y^2)(x - y)^2 = 1872$? Name these parks.

2. Sequoia and Yosemite regions in California became national parks in 1890. Find the number of national parks opened in other states in 1890 if this number is equal to the product xy, where the integers x and y satisfy either $x^4 - y^4 = 1890$ or 0 if this equation does not have any integer solutions. Name these parks.

3. The number of national parks opened in 1900 is the largest non-negative integer smaller than the largest absolute value of a solution to the equation

   $$x^{2n+1} + x^{2n} + ... + x^3 + x^2 + x - 1 = 0, n > 0.$$

   How many national parks were open in the USA in 1900? Name these parks if any.

4. How many national parks were open in the USA in 1990 if this number is the largest integer less than the largest absolute value of a root of the equation

   $$x^7 + x^6 + x^5 + x^4 + x^3 + x^2 + x - 1 = 0?$$

DOI: 10.1201/9781003229889-3

5. The number of US states that have national parks can be presented as a two-digit number with tens digit a and units digit b, where $a + \sqrt{b}$ is the largest real root of the equation $3x^5 - 5x^4 - 34x^3 - 34x^2 - 5x + 3 = 0$. Moreover, the product ab gives the number of national parks that became World Heritage Sites. How many states have national parks? How many national parks are designated as World Heritage Sites?

6. How many national parks are in Delaware, Vermont, and Wisconsin, if their numbers x, y, and z, respectively, are integers that satisfy the inequality:

$$\sqrt{x^2 + 1} + \sqrt{y^2z^2 + 1} \leq \sin\sqrt{(xy + 1)(z + 1)} + 1?$$

7. The numbers of national parks and national monuments in Colorado are presented by a and b, correspondingly, that are positive integer solutions of the following equation

$$a^2b^2 + 1 - 2a^2 - 3b^2 = 606.$$

Moreover, the number of mountain ranges in Colorado is the same as the number of national monuments. The number of national forests is the sum of a and b. The number of state parks is 14 more than the product of a and b. How many national parks, monuments, and forests are in Colorado? How many mountain ranges and state parks are in the state?

8. The area of the *largest national park*, Wrangell-St. Elias in Alaska, is 1497 times the area of the *smallest national park*, Hot Springs in, Arizona. Their total area is 33,705 km². What are the areas of these two national parks?

A.	22.5 km²
B.	32,207 km²
C.	32,208 km²
D.	33,660 km²
E.	33,682.5 km²

9. The area of the largest national park, Wrangell-St. Elias in Alaska, is 1497 times the area of the smallest national park, Hot Springs in Arizona. What is the difference between the areas of these two national parks if their total area is 33,705 km²?

10. Yellowstone National Park covers approximately 3333 sq. mi in the northwest corner of Wyoming. Some portions of the park are

in Montana and Idaho. Approximately how many square miles of the park are in Montana and Idaho if the percentages of their areas located in these states are equal to the number of hexagons (for Idaho) and the number of squares (for Montana) among 10 separate shapes constructed from 36 sticks? At least one triangle, one square, and hexagon are to be constructed. One stick is used for one side of a figure.

11. Yellowstone National Park covers approximately 3333 sq. mi in the northwest corner of Wyoming. Some portion of the park is in Montana and Idaho. Approximately how many square miles of the park are in Montana and Idaho if the percentages of their areas located in these states are equal to the number of hexagons (for Idaho) and the number of squares (for Montana) that can be constructed from 36 sticks? Use the following hints to solve the problem.

    1. 10 figures are constructed.

    2. There is at least one triangle, one square, and hexagon.

    3. One stick is used for one side of the figure.

    Explain why each hint is relevant to have a unique solution.

12. Yellowstone National Park in Wyoming, Sequoia, and Yosemite National Parks in California are the first national parks in the USA. Find their areas if the area of Yellowstone National Park is 64,007.86 acres less than 3 times the area of Yosemite National Park. Choose the smallest number of hints to solve the problem and explain why other hints are not necessary for solving the problem:

    1. Three times the area of Yellowstone National Park is 24,682.99 acres more than the sum of 7 times the area of Sequoia National Park and 5 times the area of Yosemite National Park.

    2. The area of Yosemite National Park is 46,836.15 acres less than twice the area of Sequoia National Park.

    3. The area of Yellowstone National Park is 110,844.01 acres less than twice the total area of the other two parks.

13. Rocky Mountain National Park in Colorado offers gorgeous views along hundreds of trails and scenic Trail Ridge Road. It has four visitor centers.

    The elevation of the Kawuneeche Visitor Center is 480 ft above the elevation of the Fall River Visitor Center, but 3076 ft below Alpine Visitor Center. The elevations of the Kawuneeche, Fall River, and Beaver Meadows Visitor Centers are related to as 109 to 103 to 98. What are the elevations of Alpine, Beaver Meadows, Fall River, and Kawuneeche Visitor Centers?

14. The 13,063-ft Wheeler Peak and a fantastic limestone formation of Lehman Caves are gems of Great Basin National Park, Nevada. Despite incredible trails through groves of ancient bristlecone pines and glaciers lakes, it is the least-visited national park in the nation due to its remote location.

Three campsites, Lower Lehman Creek Campground (elevation of 7300 ft), Upper Lehman Creek Campground (7750 ft), and Wheeler Peak Campground (9885 ft) are located on the Wheeler Peak Scenic Drive. What is the temperature at each campground if it is 85 F at the Lehman Caves entrance with the elevation of 6825 ft above sea level?

Following the warning sign posted on Alpine Ridge trail in Rocky Mountain National Park, Colorado, for every 1000 ft increase in elevation the temperature drops by 5 F.

15. A freestanding natural Delicate Arch is the most recognized landmark in Arches National Park of Utah. Standing 60 ft tall, Delicate Arch is depicted on a Utah license plate and a postage stamp of 1996 dedicated to the centennial anniversary of Utah admitting the Union.

Impressed with Delicate Arch, a group of 25 students decided to build its small replica from 25 lb of craft supplements. The group had at least one freshman, one sophomore, and one junior. They thought that it would be fair if each freshman used 5 lb, sophomore used 1 lb, and junior used 1/5 lb of craft supplements. How many freshmen, sophomores, and juniors were in the group? Does the problem have a unique solution? If not, what statement should be added for the uniqueness of the solution?

16. A National Historic Landmark, the Great Circle Earthworks in Ohio, contains the largest set of geometric earthworks built by North American Indian people from 100 B.C. to 500 A.D. The enclosure has a circular wall with up to 14 ft height. The outer portion of the wall is made of dark brown soil, while the inner portion is made from yellow-brown soil. A ditch at the base of the wall inside the enclosure is up to 13 ft deep. The land was transferred to the Ohio Historical Society in 1932. It is now open to the public.

What is the diameter of the Great Circle if the side of a square inscribed into the circle is $a = 600\sqrt{2}$ ft?

17. The outstanding red rock formations are the gem of Garden of the Gods, a public park in Colorado Springs. In MDCCCLXXIX, Charles Elliot Perkins purchased CDLXXX acres of land for XXII dollars per acre. Following Perkins' wish, his family gave the land to the City of Colorado Springs in MCMIX under the condition that it should be always "kept free to the entire world". It was

designated a National Natural Landmark in MCMLXXI. The original Perkins' "gift of inestimable value" forms the center of Garden of the Gods, which is currently MCCCLXVII acres. When, how much of the land, and for what price did Mr. Perkins purchase the land in Colorado? When was this land donated to the City of Colorado Springs? When did it become a National Natural Landmark? What is the area of Garden of the Gods?

18.–19. *The American Southwest is rich with fantastic scenery, unique rock formations, and deep canyons. Five national parks of Utah, Grand Canyon National Park in Arizona, Mesa Verde National Park in Colorado, Monument Valley, and Glen Canyon National Recreation Area with its Lake Powell are inside the charming Grand Circle.*

18. Cities Santa Fe, NM; Kingman, AZ; and Vernad, UT are located at the border of the Grand Circle and not in one line. The straight distances between the cities are provided in the table:

	Santa Fe, NM	Kingman, AZ
Kingman, AZ	458 mi	
Vernad, UT	384 mi	440 mi

Assuming the perfect circle shape, find the area of the Grand Circle, its diameter, and the length of the border.

19. Cities of Salida, CO; Santa Fe, NM; Kingman, AZ; and Vernal, UT are located at the border of the Grand Circle. The straight distances between the cities are provided in Figure III.1 below.

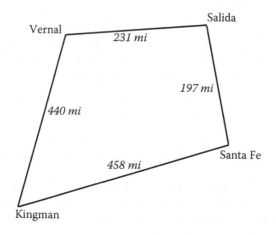

FIGURE III.1

Find the length of the circumference of the Grand Circle or explain if it is impossible to solve the problem.

Answers

1. 1
2. 0
3. 0
4. 0
5. 27, 14
6. 0, 0, 0
7. 4, 7, 11, 7, 42
8. 33,682.5 km^2, 22.5 km^2
9. D
10. 35 sq. mi, 104 sq. mi
11. 35 sq. mi, 104 sq. mi
12. 2,219,790.71, 404,051.17, 761,266.19 acres, Hints 1 and 2 or 1 and 3
13. 11,796 ft, 7840 ft, 8240 ft, 8720 ft
14. 82.6 F, 80.4 F, 69.7 F
15. No, {(1, 19, 5), (2, 13, 10), (3, 7, 15), (4, 1, 20)}.
16. 1,200 ft
17. 1879, 480, 22, 1909, 1971, 1367
18. 77,843 sq. mi, 497 mi, 1561 mi
19. Impossible

Solutions

1. 1st way. Since x and y are integers and $(x^2 - y^2)(x - y)^2 = 1872 = 2^4 \cdot 3^2 \cdot 13$, then $(x - y)^2$ takes squares from the right side leaving the rest to $(x^2 - y^2)$. Thus, the possible options are:

	1	2	3	4	5	6
$(x - y)^2$	1	2^2	2^4	3^2	$2^2 3^2$	$2^4 3^2$
$(x^2 - y^2)$	$2^4 3^2 13$	$2^2 3^2 13$	$3^2 13$	$2^4 13$	$2^2 13$	13

Instead of checking all options, let us take a closer look at the problem. Since 1872 is even, x and y are odd or even at the same time. Then their difference and the difference of their squares are even that takes away Options 1, 3, 4, and 6.

Option 5 leads to $\begin{cases} x - y = 6 \\ x^2 - y^2 = 52 \end{cases} \Rightarrow \begin{cases} x - y = 6 \\ x + y = 26/3 \end{cases} \Rightarrow$ not integer x and y.

Option 2 produces $\begin{cases} x - y = 2 \\ x^2 - y^2 = 468 \end{cases} \Rightarrow \begin{cases} x - y = 2 \\ x + y = 234 \end{cases} \Rightarrow y = 116, x = 118,$

which works. Then $\frac{x-y}{2} = 1$.

2nd way. Using algebraic identities (see Appendix I-AF) let us rewrite $(x^2 - y^2)(x - y)^2 = (x + y)(x - y)^3 = 1872 = 2^3 \cdot 2 \cdot 3^2 \cdot 13$. Since x and y are integers then $(x-y)^3$ take cubes from the right side leaving the rest to $(x+y)$. Hence $\begin{cases} x - y = 2 \\ x^2 - y^2 = 468 \end{cases}$ and $y = 116, x = 118$.

Only one park, the Yellowstone National Park, was signed into law in 1872.

2. Let us analyze $x^4 - y^4 = 1890$. Since x and y are integers and $x^4 - y^4 = (x - y)(x + y)(x^2 + y^2) = 1890$, then x and y are odd or even at the same time. Otherwise, $(x - y)$, $(x + y)$, and $(x^2 + y^2)$ are all odd and their product is odd, which contradicts that 1890 is even.

If x and y are both odd, that is $x = 2k + 1$, $y = 2n + 1$, where k and n are any integers, then both $(x - y)(x + y) = 4(k - n)(k + n + 1)$ and 1890 should be divisible at least by 4, which is not the case, $1890 = 2 \cdot 945$. The analysis is similar if x and y are both even. Thus, the equation $x^4 - y^4 = 1890$ does not have any integer solution.

Only two national parks, Sequoia and Yosemite National Parks, in California were open in 1890.

3. The solution is similar to the solution of Problem 4 of this section.

No national parks were opened in 1900.

4. The equation $p(x) = x^7 + x^6 + x^5 + x^4 + x^3 + x^2 + x - 1 = 0$ has one positive root by Descartes rule of sign (see Appendix I-PE). Since $p(0) = -1$ and $p(1) = 6$, the positive root is between 0 and 1.

This equation can have at most 6 negative real roots. Indeed, the coefficients of $p(-x) = (-x)^7 + (-x)^6 + (-x)^5 + (-x)^4 + (-x)^3 + (-x)^2 + (-x) - 1 = -x^7 + x^6 - x^5 + x^4 - x^3 + x^2 - x - 1$ change the sign 6 times. Clearly, $x = -1$ is not a solution to $p(x) = 0$. Let us consider two intervals for x: $(-1, 0)$ and $(-\infty, -1)$. Rewriting $x^7 + x^6 + x^5 + x^4 + x^3 + x^2 + x - 1 = 0$ on the first interval $(-1, 0)$ as $x^7 + x^5(x + 1) + x^3(x + 1) + x(x + 1) - 1 < 0$, one can see that there is no negative real roots on $(-1, 0)$. Considering the second interval $(-\infty, -1)$ and rewriting the equation as $x^6(x + 1) + x^4(x + 1) +$

$x^2(x + 1) + (x - 1) < 0$, one can see that there are no negative real roots on this interval either. Thus, there are no negative real roots.

Hence, the largest nonnegative integer less than the largest absolute value of a (and the only one) root is 0.

No national parks were open in 1990.

5. The equation $p(x) = 3x^5 - 5x^4 - 34x^3 - 34x^2 - 5x + 3 = 0$ is a symmetric equation of the 5th degree (check the coefficients: 3, –5, –34, –34, –5, 3). Then $x = -1$ is one of its roots. This can be verified by the synthetic division:

$$
\begin{array}{r|rrrrrr}
-1 & 3 & -5 & -34 & -34 & -5 & 3 \\
   & \downarrow & -3 & 8 & 26 & 8 & -3 \\
\hline
   & 3 & -8 & -26 & -8 & 3 & 0
\end{array}
$$

Thus, the equation can be rewritten as $(x + 1)(3x^4 - 8x^3 - 26x^2 - 8x + 3) = 0$. Let us find out whether the equation $3x^4 - 8x^3 - 26x^2 - 8x + 3 = 0$ has a solution. This equation is a symmetric equation of the 4th degree. After verifying that $x = 0$ cannot be its solution (simply by substitution 0 for x) and dividing both sides of by x^2, we obtain $3x^2 - 8x - 26 - 8\frac{1}{x} + 3\frac{1}{x^2} = 0$. This equation can be rewritten as $3t^2 - 8t - 32 = 0$ in the new variable $t = x + \frac{1}{x}$; then $t^2 = x^2 + 2 + \frac{1}{x^2}$. Solving this quadratic equation (see Appendix I-QE), we obtain $t = \frac{4 \pm 4\sqrt{7}}{3}$. Remembering the substitution $t = x + \frac{1}{x}$, we get four so-

lutions $\left\{ 2 + \sqrt{7}, \frac{-2 + \sqrt{7}}{3}, 2 - \sqrt{7}, \frac{-2 - \sqrt{7}}{3} \right\}$ to

$3x^2 - 8x - 26 - 8\frac{1}{x} + 3\frac{1}{x^2} = 0$ and five solutions $\left\{ -1, 2 + \sqrt{7}, \frac{-2 + \sqrt{7}}{3}, \right.$

$2 - \sqrt{7}, \frac{-2 - \sqrt{7}}{3} \right\}$ to the original equation $p(x) = 0$: $\left\{ -1, 2 + \sqrt{7}, \frac{-2 + \sqrt{7}}{3}, \right.$

$2 - \sqrt{7}, \frac{-2 - \sqrt{7}}{3} \right\}$.

Indeed, taking $t = \frac{4 + 4\sqrt{7}}{3}$ and using the substitution $x + \frac{1}{x} = \frac{4 + 4\sqrt{7}}{3} \Rightarrow x^2 - $

$\frac{4 + 4\sqrt{7}}{3}x + 1 = 0 \Rightarrow 3x^2 - (4 + 4\sqrt{7})x + 3 = 0 \Rightarrow x = \frac{4 + 4\sqrt{7} \pm \sqrt{(4 + 4\sqrt{7})^2 - 36}}{6}$

$= \frac{2 + 2\sqrt{7} \pm \sqrt{23 + 8\sqrt{7}}}{3} = \frac{2 + 2\sqrt{7} \pm \sqrt{16 + 2 \cdot 4\sqrt{7} + (\sqrt{7})^2}}{3} = \frac{2 + 2\sqrt{7} \pm (4 + \sqrt{7})}{3}$, which

leads to the first two roots. Its other two roots are obtained in a similar way. The largest root $2 + \sqrt{7}$ produces 27 as the number of states that have national parks and the product of 2 and 7 gives 14, which is the number of national parks designated as World Heritage Sites.

27 US states have national parks. 14 national parks are designated as World Heritage Sites.

6. Since the range of $\sin x$ is $[-1, 1]$, the maximum value of the right side $\sin \sqrt{(xy + 1)(z + 1)} + 1$ is 2. The lowest value of the left side $\sqrt{x^2 + 1} + \sqrt{y^2 z^2 + 1}$ is 2 that occurs when x, y, and z are zeroes. Thus, only $x = 0$, $y = 0$, and $z = 0$ satisfy the inequality.

Delaware, Vermont, and Wisconsin do not have any national parks. They have other national sites, such as Washington-Rochambeau Revolutionary Route National Historic Trail (Delaware), Appalachian National Scenic Trail (Vermont), Ice Age National Scenic Reserve (Wisconsin), and others.

7. $a^2 b^2 + 1 - 2a^2 - 3b^2 = 606 \Rightarrow a^2(b^2 - 2) + 1 - 3(b^2 - 2) - 6 = 606 \Rightarrow (b^2 - 2)$ $(a^2 - 3) = 611$. The number 611 can be factored as $1 \cdot 611$ or as $13 \cdot 47$ that give values to $(b^2 - 2)$ and $(a^2 - 3)$. The first product $1 \cdot 611$ does not lead to any positive integer solutions solution because neither $611 + 2$ nor $611 + 3$ are squares. The second product $13 \cdot 47$ has two cases: (1) $b^2 = 13 + 2$ and $a^2 = 47 + 3$, which are not squares and (2) $b^2 = 47 + 2$ and $a^2 = 13 + 3$, that lead to $a = 4$ and $b = 7$.

Colorado has 4 national parks, 7 national monuments, 7 mountain ranges, 11 national forests, and 42 state parks.

8. Let x km^2 be the area of the smallest national park, then the area of the largest national park is $1497x$ and from $x + 1497x = 33{,}705$ we get $x = 22.5$ km^2.

The area of the largest national park, Wrangell-St. Elias in Alaska, is 33,682.5 km^2, while 22.5 km^2 is the area of the smallest national park, Hot Springs in Arizona.

9.

A.	22.5 km^2	The area of the smallest park
B.	32,207 km^2	33,705 – 1498
C.	32,208 km^2	33,705 – 1497
D.*	33,660 km^2	Let x be the area of the smallest national park $\Rightarrow$ the area of the largest national park is $1497x$ and $x + 1497x = 33{,}705 \Rightarrow x = 22.5$; $33{,}705 - 22.5 = 33{,}660$
E.	33,682.5 km^2	The area of the largest park.

10. Let n, m, and k be the numbers of triangles, squares, and hexagons constructed from 36 sticks. Ten figures are constructed, then we get a system of two equations in three variables

$$\begin{cases} n + m + k = 10 \\ 3n + 4m + 6k = 36 \end{cases} \Rightarrow \begin{cases} k = 10 - n - m \\ 3n + 2m = 24 \end{cases} \Rightarrow \begin{cases} k = \frac{n-4}{2} \\ m = \frac{24-3n}{2} \end{cases}.$$ Because   there

should be at least one figure of each type, then $1 \leq k \leq 8$,

$$1 \leq m \leq 8 \Rightarrow \begin{cases} 1 \leq k = \frac{n-4}{2} \leq 8 \\ 1 \leq m = \frac{24-3n}{2} \leq 8 \end{cases} \Rightarrow \begin{cases} 6 \leq n \leq 20 \\ 2 \leq n \leq 7 \end{cases} \Rightarrow 6 \leq n \leq 7.$$

Furthermore, n should be even have integer m and k above. Hence, $n = 6$, $m = 3$, and $k = 1$.

Thus, $100\% - 1\% - 3\% = 96\%$, which is 3333 sq. mi, is in Wyoming. 1% of the Yellowstone National Park in Idaho is $3333 \cdot 1/96 \approx 35$ and 3% of the park or $3333 \cdot 3/96 \approx 104$ sq. mi is in Montana. 3333 sq. mi of the Yellowstone National Park is located in the northwest corner of Wyoming, 104 sq. mi of the park is located in Montana, and 35 sq. mi in Idaho.

11. If Hint 1 is omitted, then a solution is not unique because there will be one equation in three variables.

If Hint 2 is omitted, then a solution is not unique because k, m, and n can be 0, which increases the number of options for the solution. For instance, then

$$0 \leq k \leq 10,\ 0 \leq m \leq 10,\ 0 \leq n \leq 10 \Rightarrow \begin{cases} 0 \leq k = \frac{n-4}{2} \leq 10 \\ 0 \leq m = \frac{24-3n}{2} \leq 10 \end{cases} \quad 0 \leq n \leq 10^4$$

$\leq n \leq 10$ that lead to (4, 6, 0), (6, 3, 1), (8, 0, 2).

If Hint 3 is omitted, then a solution is not unique and requires consideration of various options, for instance, five sticks can be used to construct two triangles with a joint side, and so on.

3333 sq. mi of the Yellowstone National Park is located in the northwest corner of Wyoming, 104 sq. mi of the park is located in Montana, and 35 sq. mi in Idaho.

12. Let x, y, and z be the areas of Yellowstone, Sequoia, and Yosemite

National Parks correspondingly, then $\begin{cases} x\ +\ 64007.86 & = 3z \\ 3x\ -\ 24682.99\ = 7y + 5z \\ z\ +\ 46836.15 & = 2y \\ x + 110844.01 = 2y + 2z \end{cases}$ or

$$\begin{cases} -x\ +\ 3z = 64007.86 \\ 3x\ -\ 7y\ -\ 5z = 24682.99 \\ 2y\ -\ z = 46836.15 \\ -x\ +\ 2y\ +\ 2z = 110844.01 \end{cases}.$$

We obtained the system of four linear equations in three variables.

Subtracting the third equation from the fourth equations we obtain the first equation. Because the first equation is given, then only the third (Hint 2) or the fourth equation (Hint 3) is necessary in addition to Hint 1. Then

$$\begin{cases} -x & +3z = 64007.86 \\ 3x & -7y -5z = 24682.99 \\ & 2y -z = 46836.15 \end{cases}.$$

$$\underset{II+3\cdot I}{\Rightarrow} \begin{cases} -x & +3z = 64007.86 \\ & -7y +4z = 216706.57 \\ & 2y -z = 46836.15 \end{cases} \underset{II+4\cdot III}{\Rightarrow} \begin{cases} -x & +3z = 64007.86 \\ & y & = 404051.17 \\ & 2y -z = 46836.15 \end{cases}$$

$$\underset{III-2\cdot II}{\Rightarrow} \begin{cases} -x & +3z = 64007.86 \\ & y & = 404051.17 \\ & z = 761266.19 \end{cases} \underset{II-3\cdot III}{\Rightarrow} \begin{cases} x & = 2219790.71 \\ y & = 404051.17 \\ z = 761266.19 \end{cases}.$$

The areas of Yellowstone, Sequoia, and Yosemite National Parks are 2,219,790.71, 404,051.17, and 761,266.19 acres, respectively.

13. Let a, b, f, and k stand for the elevation in feet of Alpine, Beaver Meadows, Fall River, and Kawuneeche Visitor Centers. Then from $k/f/b = 109/103/98$, follows $f = 103k/109$, and $b = 98k/109$. From $k = f + 480 \Rightarrow k = 103k/109 + 480 \Rightarrow k = 8720$; $b = 98k/109 \Rightarrow b = 7840$, $f = 103k/109 = 8240$. Applying $a = k + 3076 \Rightarrow a = 11,796$.

14. $7300 - 6825 = 475$ ft $\Rightarrow \dfrac{475}{1000} \dfrac{x}{5} \Rightarrow x = \dfrac{475 \cdot 5}{1000} = 2.375 \Rightarrow 85F - 2.4F = 82.6$ F at Lower Lehman Creek Campground. Similarly, $\dfrac{(7750 - 6825) \cdot 5}{1000} = 4.625 \Rightarrow 80.4$ F at Upper Lehman Creek Campground and $\dfrac{(9885 - 6825) \cdot 5}{1000} = 15.3 \Rightarrow 69.7$ F at Wheeler Peak Campground.

15. Let $f, s, j > 0$ be the number of freshmen, sophomores, and juniors, then $\begin{cases} f + s + j = 25 \\ 5f + s + \frac{1}{5}j = 25 \end{cases} \underset{I-II}{\Rightarrow} \begin{cases} j = 5f \\ s = 25 - 6f \end{cases}$. By the second equation, f cannot be greater than 4 because then the number of sophomores s becomes negative that leaves the following options

f	s	j
1	19	5
2	13	10
3	7	15
4	1	20

Another equation is needed to find a unique solution.

16. The diagonal $d = \sqrt{a^2 + a^2} = \sqrt{2 \cdot 600^2 \cdot 2} = 1200$.

    The diameter of the Great Circle is 1200 ft.

17. Following M – 1000, D – 500, C – 100, L – 50, X – 10, V – 5, I – 1, and remembering to subtract a lower value from the higher one if it is written before the later, we can read MDCCCLXXIX as 1000 + 500 + 3 · 100 + 50 + 2 · 10 + (10 – 1) = 1879, MCMIX as 1000 + (1000 – 100) + (10 – 9) = 1909, and so on.

18. Using two formulas for the area of a triangle, $A = \sqrt{p(p-a)(p-b)(p-c)}$ and $A = \frac{abc}{4R}$, where a, b, c are the sides of a triangle and $p = 641$ mi is its half perimeter, we get $A = 77{,}843$ sq. mi, $R = \frac{abc}{4A} = 248.5$ mi. The circumference $C = 2\pi R = 1561$ mi is the length of the border.

    The area of the Grand Circle is 77,843 sq. mi, its diameter is 497 mi, and the length of the border is 1561 mi.

19. The circle can be inscribed in a quadrilateral only if the sums of opposite sides are the same. Indeed, let a circle is inscribed in the quadrilateral $ABCD$ and point P, R, S, T are joint points of the circle and quadrilateral as shown in Figure III.2. Then, by properties of the tangent lines (see Appendix I-CP), $AP = AR$; $BR = BS$, $CS = CT$, $DT = DP$ or $AP + PD + BS + SC = AR + RB + CT + TD$. In our problem 231 + 438 = 669 is not equal to 440 + 197 = 637.

    It is not possible to solve the problem. The Grand Circle is not a perfect circle.

III.2 State Parks

State parks aim to preserve beauty of the land and places of important historic events and provide recreational facilities as well. Each US state has an impressive list of state parks and beaches. According to the National Association of State Parks, more than 739 million people visit state parks each year.

1. How many state parks are in the USA if this number can be presented by five consecutive digits in the increasing order except for the first two digits that are interchanged to have a five-digit number.

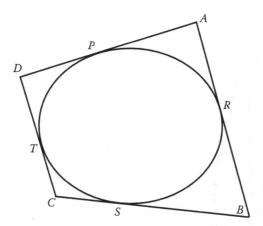

FIGURE III.2

2. According to the National Association of State Parks, state parks serve x times more visitors compared to the National Park System with only y% of the National Park System acreage. The numbers x and y are the coordinates (x, y) of the center of the circle $4x^2 + 4y^2 + 25 = 20x + 128y$. How many more visitors are in state parks than in national parks? What percentage of national park area are state parks?

3. According to the National Association of State Parks, the numbers of cabins and campsites (in thousand units) in state parks are the numerator and denominator of $\dfrac{2^0}{2^1} - \dfrac{2^1}{2^2 + \dfrac{2^0 + 2^1}{2^3 + \dfrac{2^0}{2^0 + 2^1}}}$ simplified to an unreducible fraction. How many cabins and campsites are in state parks?

4. Niagara Falls State Park in New York is the oldest state park in the USA. It was established in the year which is the least common multiple of 145 and 377. When was Niagara Falls State Park established?

5. Alaska has the largest park acreage followed by California (second) and New York (third). The total area of parks in these three states take 45% of all US state park land.

The area of state parks in Alaska is 0.5 mln acres more than the areas of state parks in California and New York together. The area of state parks in Alaska is 0.4 mln acres more than twice the area of state parks in California, but 0.8 mln acres less than triple the area of state parks in New York. What are the areas of state parks in Alaska, California, and New York?

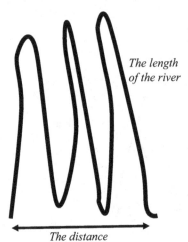

The length of the river

The distance

FIGURE III.3

6. The San Juan River has carved a deep canyon and makes a series of tight turns, called goosenecks, which can be observed in the Goosenecks State Park, Utah. The river flows several miles over quite a small distance, see Figure III.3.

   The distance that the river takes in the park is four times less than its length. Their geometric mean is 3. What is the length of the San Juan River in the Goosenecks State Park?

7. An astonishing view of beautiful waterfalls are among main attractions of the 105-acre Palouse Falls State Park, Washington. The Palouse Falls of Palouse River consist of an upper fall and a lower fall.

   The Palouse Falls drop x ft, where x is defied by $\sqrt{\sqrt{\sqrt{2x - 140}}} = 2$. How tall are the falls?

8.–9. *A U-shaped scenic state turnpike highway M22 in Michigan, Michigan Heritage Route, is a part of the Lake Michigan Circle Tour. It follows the lake shoreline and runs through Sleeping Bear Dunes National Lakeshore, Leelanau and Traverse City State Parks.*

8. The length in miles of Michigan Heritage Route is the lowest three-digit number that can be presented as a sum of squares of two positive integers or as a sum of squares of other three positive integers. The sum of the two integers that form the first sum of squares is equal to the sum of the three integers of the second sum. Both sets of integers (from the first and second sums) form two arithmetic sequences that have different first terms but the same

difference of the arithmetic sequences. The integers from the first sum are two consecutive terms. The integers from the second sum are the first, second, and fourth terms of the sequence. What is the length of Michigan Heritage Route?

9. Find the length of M22 in miles if its length is the lowest three-digit number that can be presented as a sum of squares of two integers or as a sum of squares of a different set of three integers. Use the following hints:

   1. The sum of the two integers that form the first sum of squares is equal to the sum of the three integers of the second sum.
   2. The largest integer from the set of five integers is equal to the sum of the largest integer of the first sum and the smallest integer of the second sum.
   3. The largest integer equals the sum of the smallest integer of the first sum and the middle integer of the second sum.
   4. All integers are positive.

   Are all hints needed?

10. Lincoln Memorial, Carter G. Woodson Home National Historic Site, Star-Spangled Banner National Historic Trail, and Chesapeake and Ohio Canal National Historical Park are among the national historic landmarks of Washington, D.C.

    The numbers of memorials and national historic sites in Washington, D.C., are the largest and lowest integer values of x and the numbers of national historic trails and national historic parks are the largest and lowest integer values of y correspondingly that satisfy the equation: $xy^2 + 15x + 12y^2 = 252$. The number of state parks is one less than the number of historic parks. How many state parks, memorials, national historic sites, national historic trails, and national historic parks are in Washington D.C.?

Answers

1. 10,234
2. 2½, 16%
3. 9 thousand, 218 thousand
4. 1885
5. 3.4 mln acres, 1.5 mln acres, 1.4 mln acres
6. 6 mi

7. 198 ft

8. 117 mi

9. 117 mi, no

10. 0, 15, 6, 3, 1

Solutions

1. The first digit of 0 violates the statement that the number of parks is a five-digit number. Thus, from 01234 we get 10,234 state parks. There were 10,234 state parks in the USA in 2014.

2. Rewriting $4x^2 + 4y^2 + 25 = 20x + 128y$ as $4(x^2 - 2 \cdot 5/2 \cdot x + 25/4) + 4(y^2 - 2 \cdot 16y + 16^2) - 4 \cdot 16^2 = 0$ leads to $(x - 5/2)^2 + (y - 16)^2 = 16^2$. Thus, the center of the circle is at the point $(5/2, 16)$.

   State parks welcome 2.5 more visitors than national parks though their area is 16% of the area of national park areas.

3. The expression is simplified to $\dfrac{2^0}{2^1} - \dfrac{2^1}{2^2 + \frac{2^0 + 2^1}{2^3 + \frac{2^0}{2^0 + 2^1}}} = \dfrac{1}{2} - \dfrac{2}{4 + \frac{9}{25}} = \dfrac{1}{2} - \dfrac{2 \cdot 25}{109} = \dfrac{9}{218}$.

   There are 9 thousand cabins and 218 thousand campsites in US state parks.

4. Prime factorization of 145 is $5 \cdot 29$ and of 377 is $13 \cdot 29$. The least common multiple of 145 and 377 is $5 \cdot 29 \cdot 13 = 1885$, which is the year of establishing the first state park in the USA.

   The first state park in the USA, Niagara Falls State Park in New York, was opened in 1885.

5. Let n, c, and a be areas in mln acres of state parks in New York, California, and Alaska. Then
   $$\begin{cases} a = c + n + 0.5 \\ a = 2c + 0.4 \\ a = 3n - 0.8 \end{cases}$$
   . Subtracting the second equation from the first one and the third equation from the second one, we get $\begin{cases} 3n - 2c = 1.2 \\ c - n = 0.1 \end{cases}$ that leads to $n = 1.4$, $c = 1.5$, and $a = 3.4$.

   State parks in New York take 1.4 mln acres, 1.5 mln acres of state parks are in California, and 3.4 mln acres of state parks are in Alaska.

6. Let x and y be the length and the distance, then $x = 4y$, $\sqrt{xy} = 3 \Rightarrow \sqrt{4y^2} = 3 \Rightarrow y = \frac{3}{2}, x = 6$.

The San Juan River in the Goosenecks State Park is 6 mi long.

7. The Domain of $\sqrt{\sqrt{\sqrt{2x - 140}}}$ is $x \geq 70$. Raising to the 8th power both sides of $\sqrt{\sqrt{\sqrt{2x - 140}}} = 2$ or $(2x - 140)^{\frac{1}{8}} = 2$, we get $2x - 140 = 2^8 \Rightarrow x = 198$.

The Palouse Falls are 198 ft.

8. Let l be the length, a and c be the first terms of the two sums, and d be the difference of both arithmetic sequences. From $a + (a + d) = c + (c + d) + (c + 3d)$ follows $2a = 3(c + d)$, which after substitution to $l = a^2 + (a + d)^2 = c^2 + (c + d)^2 + (c + 3d)^2$ leads to $3c^2 + 8cd - 3d^2 = 0$. Solving the last quadratic equation with respect to c we get $c = \frac{-4d \pm \sqrt{16d^2 + 9d^2}}{3}$ or $c = d/3$ and $c = -3d$. If $c = -3d$, then the last term that forms the second sum, which is the fourth term of the second arithmetic sequence, will be 0, or will be nonpositive. Therefore, we will consider $c = d/3$. In order for c to be integer, d should be divisible by 3. Checking the first option $d = 3$ leads to $c = 1$ and $1^2 + 4^2 + 10^2 = 6^2 + 9^2 = 117$.

The Michigan Heritage Route is 117 mi.

9. Let l be the length and $a, b, c, d,$ and $e, a > b, c > d > e$ be five integers. Then $l = a^2 + b^2 = c^2 + d^2 + e^2, a + b = c + d + e$ (Hint 1), $c > a$ and $c = a + e$ (Hint 2), $c = b + d$ (Hint 3).

Let us start with the lowest integer $e = 1$, then $\begin{cases} a = 2d + 1 \\ b = d + 2 \\ c = 2d + 2 \end{cases}$.

If $d = 1$, then $3^2 + 3^2 = 4^2 + 1^2 + 1^2 = 18$. If $d = 2$ then $5^2 + 4^2 = 6^2 + 2^2 + 1^2 = 41$.
If $d = 3$, then $7^2 + 5^2 = 8^2 + 3^2 + 1^2 = 74$.
If $d = 4$, then $9^2 + 6^2 = 10^2 + 4^2 + 1^2 = 117$. The lowest three-digit number is 117.

From $a^2 + b^2 = c^2 + d^2 + e^2, a + b = c + d + e, c = a + e, c = b + d$ follows that $3c = 2a + 2b$. Thus, not all hints are needed.

The Michigan Heritage Route is 117 mi.

10. The equation $xy^2 + 15x + 12y^2 = 252 \Leftrightarrow xy^2 + 15x + 12y^2 + 12\cdot15 = 252 + 12\cdot15 \Leftrightarrow (x + 12)(y^2 + 15) = 432$. Since x and y are positive integers, then $x + 12 > 12$ and $y^2 + 15 > 15$ and 432 should be presented as a product of two positive integers not less than 12 and 15, or as

   a. $432 = 27 \cdot 16$, then $x = 15$, $y = 1$ ($y^2 = 1$), it works.

   b. $432 = 24 \cdot 18$, then $x = 12$, $y^2 = 3$, it does not work, because y is not integer.

   c. $432 = 16 \cdot 27$, then $x = 4$, $y^2 = 12$, it does not work, because y is not integer.

   d. $432 = 18 \cdot 24$, then $x = 6$, $y = 3$ ($y^2 = 9$), it works.

Washington, D.C. has 15 memorials, 6 national historic sites, 3 national historic trails, and one national historic park.

III.3 National Trails

National trails are all over the USA passing through forest and wild lands, charming towns and lovely farms, wooden hills and fertile meadows. Although they are different in length and significance, all of them are tremendously scenic. The Appalachian National Scenic Trail and Pacific Crest Trail are the first trails to become National Scenic Trails under the National Trails System Act of 1968.

1. Stretching from Georgia to Maine, the *Appalachian National Scenic Trail* is among the longest (if not the longest) hiking-only trails in the world. It was completed in 1937 and passes through 14 states: Georgia, North Carolina, Tennessee, Virginia, West Virginia, Maryland, Pennsylvania, New Jersey, New York, Connecticut, Massachusetts, Vermont, New Hampshire, and Maine.

   The second National Scenic Trails constructed in 1968, the 2659-mi *Pacific Crest Trail*, runs from the US border with Mexico in California through Oregon to the US–Canada border in Washington and ranges in elevation from just above sea level to 13,153 in the Sierra Nevada. The route passes through 25 national forests and 7 national parks.

   It takes about six months to complete the 3100-mi *Continental Divide National Scenic Trail* between Mexico and Canada that goes through Montana, Idaho, Wyoming, Colorado, and New Mexico. It became a National Scenic Trail in 1978.

   The Appalachian Trail, the Continental Divide National Scenic Trail, and the Pacific Crest Trail form the Triple Crown of long-distance hiking in the USA.

How many National Scenic Trails are in the USA if this number is defined by x that satisfies the equation

$$(-x) \cdot |-x| = -121?$$

Have you walked any of them?

2. National Millennium Trails are long distance trails designated by the White House Millennium Council on June 26, 2000 to reflect America's history and culture and preserve the nation's gorgeous past.

How many National Millennium Trails are in the USA, if this number is the smallest number with positive factors that satisfy the equation

$$\left(\frac{x}{2} - 1\right)^{x - 2^{2^2}} = \left(\frac{x}{2} - 1\right)^{2^{2^2} - x}?$$

3.–4. *The Great Western Trail is a unique north-south long-distance multiple use route which passes through Rockies at five western US states, Arizona, Idaho, Montana, Utah, and Wyoming.*

3. The number formed by the first two digits of the four-digit length of the Great Western Trail in miles is related to the number formed by its last two digits as 4 to 5. What is the ratio between the sum of these two two-digit numbers and the trail length? What is the length of the trail? Is it possible to answer these questions using the given data?

4. The number formed by the first two digits of the four-digit length of the Great Western Trail in miles is related to the number formed by the last two digits as 4 to 5. What is the length of the trail if the absolute value of the difference between the two two-digit numbers is 11?

5. The *Juan Bautista de Anza National Historic Trail* is a National Millennium Trail and the United States National Historic Trail. Starting at the US–Mexico border in Arizona, it runs through the California desert and coastal areas to San Francisco marking the route of Spanish exploration led by Juan Bautista de Anza (1736–1788).

Another National Millennium Trail, the *Hatfield–McCoy Trails* in West Virginia and Kentucky, is a trail system popular for its gorgeous views of mountain ridges and off-highway vehicle trails. The trail system is named after two families, the Hatfields and McCoys.

Development of the *East Coast Greenway* began in 1991 to create an urban greenway/rail trail to connect 15 American states on the Atlantic coast from Maine to Florida. About 30% of this National Millennium Trail has been completed.

A north-south long-distance National Millennium Trail, the *Great Western Trail*, stretches from Canada to Mexico through Arizona, Utah, Idaho, Wyoming, and Montana.

The longest of National Millennium Trails, the *North Country National Scenic Trail*, runs from eastern New York to central North Dakota passing seven states.

The lengths (in miles) of the Hatfield–McCoy and North Country National Scenic Trails are the first and ninth terms of the following sequence

$$1000, 1100, 1300, 1600, 2000, \ldots.$$

The length of the Juan Bautista de Anza National Historic Trail is the average between the second and third terms of the sequence. The lengths of the East Coast Greenway Trail and the Great Western Trail are 100 mi less than the seventh and ninth terms of the sequence. Find the length of all trails. Write the general and recurrence formulas of the sequence.

6. The lengths of the *East Coast Greenway* Trail from Maine to Florida, the skiing trail in Vermont *Catamount*, and the *Art Loeb* Trail in Pisgah National Forest, North Carolina, form a geometric sequence with the common ratio of 1/10. The difference between the first two terms is 1/10 of the third term cubed. How long are the East Coast Greenway, Catamount, and Art Loeb Trails?

7. The *Bonneville Shoreline* circles the ancient Lake Bonneville in Utah, while the *Great Allegheny Passage* is a rail trail between Maryland and Pennsylvania. The *Lone Star Hiking Trail* is in Sam Houston National Forest in Texas, while the *Mark Hatfield Memorial Trail* is a wildness trail in Oregon. The *Tahoe-Yosemite Trail* is located in Lake Tahoe and Yosemite National Park, California. The *West Rim Trail* is over the western edge of Pine Creek Gorge in Pennsylvania.

The lengths of six trails, the Bonneville Shoreline, the Great Allegheny Passage, the Lone Star Hiking Trai, the Mark Hatfield Memorial Trail, the Tahoe-Yosemite Trail, the West Rim Trail, are the first six terms of an arithmetic sequence. The trails are listed in the alphabetical order, NOT as consecutive terms of the arithmetic sequence. The Lone Star

Hiking Trail is 60 mi shorter than Tahoe-Yosemite Trail. The Great Allegheny Passage is 90 mi longer than Mark Hatfield Memorial Trail, which is longer than the West Rim Trail. The length of Tahoe-Yosemite Trail is the same as the lengths of Lone Star Hiking and Mark Hatfield Memorial Trails taken together. The length of each trail is rounded to an integer. What are the lengths of all trails?

8. The *Wonderland Trail* encircles Mount Rainier in the Mount Rainier National Park, Washington. It is a gorgeous though strenuous hike with frequent change in elevation.

   The units digit of a two-digit length of the trail in miles is the square root of the tens digit. The difference between these digits is twice one of the digits. How long is the Wonderland Trail?

9.–10. *The John Muir Trail in the Sierra Nevada mountain range in California is often named the America's most famous trail. This long-distance trail begins at the Happy Isles in Yosemite Valleys and passes through Yosemite, Kings Canyon, Sequoia National Parks, Inyo National Forest, spectacular alpine fields, lakes, and falls before it ends on the summit of Mount Whitney, the highest point in the contiguous USA.*

9. Except for the first 1/30 of the John Muir Trail, its elevation never falls below 7000 ft. About 35% of the trail, including the last 1/7 of the trail, lie above 10,000 ft. The elevation of only 129.5 mi of the trail is between 7000 and 10,000 ft. How long is the John Muir Trail?

10. The number of mountain passes over 11,000 ft along the John Muir Trail is the value that the function

$$f(x) = \frac{\sqrt[3]{32x} + \sqrt[4]{16x} + \sqrt[8]{8x} + \sqrt[16]{2x} + \sqrt[32]{x} + \sqrt[4]{64x^2} + \sqrt[8]{32x^2} + \sqrt[16]{16x^2} + \sqrt[32]{8x^2} + \sqrt[64]{4x^2}}{\sqrt[3]{2x} + \sqrt[4]{4x} + \sqrt[8]{8x} + \sqrt[16]{16x} + \sqrt[32]{32x} + \sqrt[64]{64x}}$$

   approaches as x approaches infinity ∞. How many passes over 11,000 ft does the John Muir Trail cross?

11. An awesome walking and biking trail, the *Wild Azalea Trail* in Louisiana, has been designated as a National Recreation Trail because of its outstanding beauty.

   The Wild Azalea Trail is 31-mi long. A walker began his hike at its starting point at Valentine Lake toward his car that he left at the trail terminal point at Woodworth. After 7 hours and 12 minutes, the walker remembered that he forgot the key in a car at Valentine Lake and called his friend, a cyclist, to bring it. The walker did not want to wait for his friend to come and continued walking. The cyclist left Valentine Lake immediately after the phone call and, biking at 10 mph, brought the key to the walker. Then the cyclist turned back and arrived at Valentine Lake when the walker was 1 mi away from the end of the trail at Woodworth. What was the average speed of the walker?

12. Olga and Travis live at two different end points of the 26-mi *Fox River State Recreational Trail* in Wisconsin. They usually leave their houses at the same time and meet at the middle. One day Travis left his house 1 hour after Olga. He walked with his usual average speed for 5 mi, rested for 1 hour, and then walked 0.5 mph faster. Walking with her usual speed, Olga walked 4 mi more than Travis before they met. How fast do Olga and Travis walk?

13. There are many scenic long-distance trails in the USA. Some of them start at one point and end at another, while the *Timberline Trail* in Oregon and the *Cranberry Lake Trail* in New York form a loop. The first one circles the Mount Hood, while the second trail follows the shores of Cranberry Lake.

    The total length of both trails is 90 mi. To improve the quality of the trails, the same number of benches is planned to put along each trail. The distance between the benches is the same at each trail. If the distance between the benches along the Timberline Trail is used to set up benches in the Cranberry Lake Trail, then 25 benches are needed. If the distance between the benches along the Cranberry Lake Trail is used to install benches in the Timberline Trail, then 16 benches are needed. What is the length of the Timberline and Cranberry Lake Trails?

14. The longest continuous single-track mountain biking trail in America, the *Maah Daah Hey Trail*, goes from Sully Creek State Park to the North Unit of Theodore Roosevelt National Park connecting the northern and southern portions of Theodore Roosevelt National Park. *Maah Daah Hey* for Mandan Indian means *an area that has been or will be around for a long time*.

    Two cyclists biked the trail. The distance between them was 16 mi at one moment. If the two cyclists biked in the same direction, then the second cyclist would have caught the first one in 4 hours. If they biked toward each other then they would meet in 48 minutes. What is the length of the Maah Daah Hey Trail if the second cyclist needs 4 hours less to bike the entire trail than the first one?

15. The *Race Across America* is an ultramarathon bicycle race across the USA. It started in 1982 and, though it varies each year, the direction has always been from the west coast of the USA to its east coast.

The length of the Race Across America is 3000 mi. A couple decided to drive one of its routes. A husband drove a half of the distance. Then his wife drove until a half of what she had driven was left. Then the husband took a wheel. What portion of their total route was left for husband? Do you need to know the length of the Race Across America?

A. ¼

B. 1/6

C. ½

D. 1/3

E. 3/4

Answers

1. 11

2. 16

3. 1 to 45, no

4. 4455 mi

5. 1000 mi, 1200 mi, 3000 mi, 4500 mi, 4600 mi

6. 3000 mi, 300 mi, 30 mi

7. 90, 150, 120, 60, 180, 30

8. 93 mi

9. 210 mi

10. 6

11. 2.5 mph

12. 2.5 mph

13. 40 mi, 50 mi

14. 96 mi

15. B, no

Solutions

1. Since $|-x| = |x| \geq 0$ and $(-x)\cdot|-x| = -121$, then $x \geq 0$. Hence, $x^2 = 121$ and $x = 11$.

   There are 11 National Millennium Trails in the USA.

2. The equation is defined if $x \geq 2$. Because the bases in both sides are the same and 2^{2^2} is 16, then $x - 16 = 16 - x$ and $x = 16$. Moreover, the equation is also valid if $x/2 - 1 = 0$ or $x/2 - 1 = 1$, i.e., $x = 2$ or $x = 4$, because 1 raised to any power is 1 and 0 raised to any positive power is 0. Thus, $\{2, 4, 16\}$ satisfy the equation. All of them are factors of 16. They are also factors of other numbers, e.g., 32, 48, but 16 is the smallest.

   There are 16 National Scenic Trails in the USA.

3. Let x and y be the numbers formed by the first and last two digits of the four-digit length in miles, then $\frac{x}{y} = \frac{4}{5}$ and the trail length is $100x + y$. The ratio between the sum of the two two-digit numbers and the trail length is

$$\frac{x+y}{100x+y} = \frac{x}{100x+y} + \frac{y}{100x+y} = \frac{1}{100 + \frac{y}{x}} + \frac{1}{100\cdot\frac{x}{y} + 1} = \frac{1}{100 + \frac{5}{4}}$$

$$+ \frac{1}{100\cdot\frac{4}{5} + 1} = \frac{1}{45}.$$

   Provided information is not enough to find the length of the Great Western Trail.

4. Let x and y be the numbers formed by the first and last two digits of the four-digit length in miles, then $\frac{x}{y} = \frac{4}{5}$, $x < y$, and $|x - y| = y - x = 11$. Thus, $5x = 44 + 4x$ and $x = 44$.

   The length of the Great Western Trail is 4455 mi.

5. Let us find the pattern in the sequence 1000, 1100, 1300, 1600, 2000, ... The differences between the terms

$$100, 200, 300, 400, \ldots,$$

   form the arithmetic sequence with the first term of 100 and the common difference of 100. Its each term is presented as $100 + 100(n - 1) = 100n$. Hence, each term of the original sequence 1000, 1100, 1300, 1600, 2000, ... can be defined by the recursive formula as

$$a_n = a_{n-1} + 100(n - 1), \, a_1 = 1000,$$

or by the general formula as

$$a_n = 1000 + 50(n - 1)n, \, n = 1, 2, 3, \dots.$$

The term $50(n - 1)n$ is the sum of n-terms of the arithmetic sequence 100, 200, 300, 400, ... , needed to obtain a_n.

Then the first, sixth, seventh, and ninth terms are found as 1000, 2500, 3100, 4600. The average between the second and third terms is 1200.

The lengths of the Hatfield–McCoy Trail is 1000 mi, the East Coast Greenway is 3000 mi, Great Western Trail is 4500 mi, and North Country National Scenic Trails is 4600 mi, the Juan Bautista de Anza National Historic Trail is 1200 mi.

6. Let x be the third term of the geometric sequence, then $10x$ and $100x$ are its second and first terms and $100x - 10x = x^3/10 \Rightarrow$ $x(900 - x^2) = 0$, which gives $x = 0$, $x = 30$, $x = -30$.

The Art Loeb Trail is 30 mi. Catamount Trail and East Coast Greenway are 300 and 3000 mi.

7. It is obvious that the Mark Hatfield Memorial Trail is 60 mi (compare: *the Lone Star Hiking Trail is 60 mi shorter than the Tahoe-Yosemite Trail with the length of the Tahoe-Yosemite Trail is the same as the lengths of the Lone Star Hiking and the Mark Hatfield Memorial Trails taken together*). Then the Great Allegheny Passage Trail is 150 mi long (from *the Great Allegheny Passage is 90 mi longer than Mark Hatfield Memorial Trail*)

The difference of the arithmetic sequence should be a factor of 30 = GCD(60, 150). On the other side it should not be less than 18 = 90/5, which occurs only if the Great Allegheny Passage Trail and the Mark Hatfield Memorial Trail are the last and first terms of the sequence. The only number that satisfies those two conditions is 30. Knowing that the *Mark Hatfield Memorial Trail is longer than the West Rim Trail*, we get that the West Rim Trail is 30 mi (because another option of $0 = 60 - 2 \cdot 30$ does not work). Thus, the six terms of the arithmetic sequence are 30, 60, 90, 120, 150, 180. Underlined terms have been assigned. The only "unassigned" terms with the difference of 60 are 120 and 180 (*The Lone Star Hiking Trail is 60 mi shorter than the Tahoe-Yosemite Trail*).

The Lone Star Hiking Trail is 120 mi, the Tahoe-Yosemite Trail is 180 mi long, and the Bonneville Shoreline Trail is 90 mi.

8. The two-digit numbers that have the units digit as the square root of the tens digit are 11, 42, 93. Only 93 satisfies the condition that the difference between the two digits is twice one of the digits: $9 - 3 = 6 = 2 \cdot 3$.

The Wonderland Trail is 93 mi.

9. Let x mi be length of the John Muir Trail, then $(\frac{1}{30}x + 0.35x)$ miles are below 7000 and $x - (\frac{1}{30}x + 0.35x) = 129.5 \Rightarrow \frac{37}{60}x = 129.5 \Rightarrow x = 210$.

The John Muir Trail is 210 mi.

10. Let us divide the numerator and denominator by $\sqrt{x} = \sqrt[4]{x^2}$ and then simplify

$$= \frac{\sqrt[2]{32x + \sqrt[4]{16x + \sqrt[8]{8x + \sqrt[16]{2x + \sqrt[32]{x}}}}} + \sqrt[4]{64x^2 + \sqrt[8]{32x^2 + \sqrt[16]{16x^2 + \sqrt[32]{8x^2 + \sqrt[64]{4x^2}}}}}}{\sqrt[2]{2x} + \sqrt[4]{4x} + \sqrt[8]{8x} + \sqrt[16]{16x} + \sqrt[32]{32x} + \sqrt[64]{64x}}$$

$$= \frac{\dfrac{\sqrt[2]{\dfrac{32x + \sqrt[4]{16x + \sqrt[8]{8x + \sqrt[16]{2x + \sqrt[32]{x}}}}}{x}} + \sqrt[4]{\dfrac{64x^2 + \sqrt[8]{32x^2 + \sqrt[16]{16x^2 + \sqrt[32]{8x^2 + \sqrt[64]{4x^2}}}}}{x^2}}}{\dfrac{\sqrt[2]{2x} + \sqrt[4]{4x} + \sqrt[8]{8x} + \sqrt[16]{16x} + \sqrt[32]{32x} + \sqrt[64]{64x}}{\sqrt{x}}}}{}$$

$$= \frac{\sqrt[2]{32 + \sqrt[4]{\dfrac{16}{x} + \dfrac{\sqrt[8]{8x + \sqrt[16]{2x + \sqrt[32]{x}}}}{x^2}}} + \sqrt[4]{64 + \sqrt[8]{\dfrac{32}{x^2} + \dfrac{\sqrt[16]{16x^2 + \sqrt[32]{8x^2 + \sqrt[64]{4x^2}}}}{x^4}}}}{\sqrt[2]{2} + \sqrt[4]{\dfrac{4}{x}} + \dfrac{\sqrt[8]{8x + \sqrt[16]{16x + \sqrt[32]{32x + \sqrt[64]{64x}}}}}{\sqrt{x}}},$$

which approaches $\dfrac{\sqrt[2]{32 + } + \sqrt[4]{64}}{\sqrt[2]{2}} = \dfrac{4\sqrt{2} + 2\sqrt{2}}{\sqrt{2}} = 6$ for large x because then $\dfrac{16}{x}, \sqrt[4]{\dfrac{4}{x}}$, and other terms approach 0.

Six passes are Donohue Pass, Silver Pass, Selden Pass, Muir Pass, Mather Pass, Pinchot Pass, Glenn Pass, and Forester Pass.

11. 7 hours 12 minutes is 7.2 hours. Let u mph and t hours be speed of the walker and the time it took the cyclist to catch the walker (see Figure III.4), then $\begin{cases} 7.2u + tu = 10t \\ \frac{31-1}{u} = 7.2 + t + t \end{cases}$. The first equation represents the

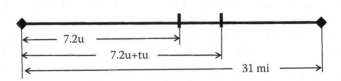

FIGURE III.4

distance the walker and cyclist took before they met. The left hand
of the second equation shows the time for the walker to cover 30
mi (1 mi left to the end of the trail), which is 7.2 hours longer than
the time the cyclist was on the trail (toward the meeting point and
the same distance back).

From $7.2u + tu = 10t \Rightarrow t = \frac{7.2u}{10-u}$ and $\frac{30}{u} = 7.2 + \frac{7.2u}{10-u} + \frac{7.2u}{10-u} \Rightarrow 7.2$
$u^2 + 102u - 300 = 0 \Rightarrow u = 2.5; u = -16.7.$

The average speed of the walker was 2.5 mph.

12. Because Olga and Travis meet at the middle, their usual average
speed is the same. Let us denote it by u. That day Olga walked
4 mi more than Travis, i.e., $(26 + 4)/2 = 15$ mi. Olga walked
11 mi. The time Travis spent before meeting her was
$1 + \frac{5}{u} + 1 + \frac{6}{u+.5} = \frac{15}{u} \Rightarrow 2u^2 - 3u - 5 = 0 \Rightarrow u = 5/2$ or $-1/2$.

Olga and Travis walk 2.5 mph.

13. Let $C, c, T,$ and t be the length of a trail and distance between the
benches along the Cranberry Lake and Timberline Trails. From
$\frac{C}{t} = 25 \Rightarrow t = \frac{C}{25}$ and $\frac{T}{c} = 16 \Rightarrow c = \frac{T}{16}$, and substituting them to
$\frac{C}{c} = \frac{T}{t}$ (both loops has the same number of benches), we get $4C = 5T$,
which in combination with $C + T = 90$ leads to $C = 50$ and $T = 40$.

Benches along the Cranberry Lake Trail are 50 ft apart and 40 ft
along Timberline Trail.

14. 48 minutes = 0.8 hours. Let u and v mph be the speed of the first and
second cyclists (see Figure III.5) then $\begin{cases} 16 + 4v = 4u \\ 16 - 0.8v = 0.8u \end{cases} \Rightarrow \begin{array}{l} v = 8 \\ u = 12 \end{array}$. If
x mi is the length of the Maah Daah Hey Trail then
$\frac{x}{8} - \frac{x}{12} = 4 \Rightarrow x = 96.$

The Maah Daah Hey Trail is 96 mi.

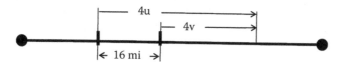

FIGURE III.5

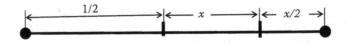

FIGURE III.6

15. Let 1 be entire route, x be the portion the wife drove, then by Figure III.6, $\frac{1}{2} = x + \frac{x}{2} \Rightarrow x = \frac{1}{3}$. So, $\frac{1}{2}x = \frac{1}{6}$.

The husband drove the final 1/6 of their trip.

IV

The US Highways

IV.1 Trips to National Parks

More and more visitors come to the US national parks each year. They take well-developed trails to enjoy breathtaking sceneries, watch diverse wildlife, and learn about historic events. Tourists often visit several national parks during one trip.

1.–4. *One of the most famous parks in the nation, Grand Canyon National Park in Arizona, was established in 1919 to preserve the exceptional canyon carved by the Colorado River and unique combinations of geologic colors. The Grand Canyon is 8 mi wide and a mile deep.*

The Glacier National Park established in 1910 in Rocky Mountains, Montana, is often called the Crown of the Continent. Its hiking and cycling trails offer scenic views of glacier-carved peaks, lakes, and valleys.

1. A family visited Grand Canyon and Glacier National Parks. It took them 4 days to drive from Grand Canyon to Glacier National Park. During the first day they covered 1/35 of the driving distance between the parks. They doubled that distance and drove 2 mi more on the second day of the trip. The family tripled the distance of the previous day and covered 4 mi more on the third day. On the fourth day they quadrupled the distance of the third day and drove 8 mi more to reach Glacier National Park. What is the driving distance between Grand Canyon and Glacier National Parks?

2. Two cars left the North Rim of *Grand Canyon National Park* for the South Rim at the same time. The faster car drove at the same speed throughout the entire route. The second car kept the same speed for the first 2 hours and then spent 20 minutes less each hour to cover the same distance as during the previous hour. The distance between two cars was 36 mi after 2 hours and 32 mi after 4 hours. How fast the cars were driving when they left the North Rim? Find the driving distance between the North and South Rims.

3. Two cars left the North Rim of *Grand Canyon National Park* toward

the South Rim at the same time. The first car drove the first three hours at the same speed, made a 30-minute stop, and kept on driving at that speed for another hour before arriving at the South Rim. The second car drove at the constant speed for the first 2 hours. The driver spent 20 minutes less during the third and fourth hours to cover the same distance as in the previous hour. The driver finished the trip at the average speed of the fourth hour. The two cars arrived at the South Rim at the same time. How fast the cars were driving when they left the North Rim if the distance between the two cars was 36 mi after 2 hours? What is the driving distance between the North and South Rims?

4. Two cars left the North Rim of *Grand Canyon National Park* toward the South Rim. The first (faster) car drove at the same speed the entire route. The second did not change its speed for the first 2 hours and then spent 20 minutes less each hour to cover the same distance as during the previous hour. What was the speed of both cars when they left the North Rim? Use the smallest number of hints listed below to solve the problem.

   A. The distance between the North Rim and the South Rim is 10–15 mi.

   B. The driving distance between the North Rim and the South Rim is 216 mi.

   C. The distance between the two cars was 36 mi after 2 hours.

   D. The distance between the two cars was 32 mi after 4 hours.

   E. It took the first car 4 hours to drive from the North Rim to the South Rim.

   F. The first car arrived at the South Rim 22 minutes and 30 seconds before the second car.

   G. The average speed of the second car was $49\frac{13}{35}$ mph.

   H. The average speed of the second car during the fourth hour was 10 mph more than the speed of the first car.

5. A *nautical mile* is a distance unit used by sea and air navigators. A *straight* or air distance between *Glacier* (Montana) and *Theodor Roosevelt* (North Dakota) *National Parks* is 554.285 km or 299.29 nautical mi, between *Yellowstone* (Wyoming) and Theodor Roosevelt National Parks is 382.28 mi or 615.22 km, and between Yellowstone and Glacier National Parks is 290.97 nautical mi. What is the air distance between Yellowstone and Glacier National Parks in miles?

6. Authentic ruins can be explored at different places. They are well preserved at *Chaco Culture National Historic Park*, New Mexico, and

Mesa Verde National Park, Colorado.

The distance between Chaco Culture National Historic Park and Mesa Verde National Park is 144 mi. One car left Chaco Culture Park toward Mesa Verde. The second car left Mesa Verde Park for Chaco Culture Park 1 hour and 15 minutes later. The cars met 1 hour after the second car left Mesa Verde. If the first car left 30 minutes after the second car, then they would meet 1 hour and 30 minutes after it left Chaco Culture National Historic Park. How fast both cars were driving?

7. *Petrified Forest National Park* in Arizona is famous for its Painted Desert and fossilized prehistoric forest. It became a national monument in 1906 and a national park in 1962. The current area of the park is about 146 sq. mi. Nine overlook points of the *Park Scenic Drive* provide an astonishing view of petrified forest and petroglyphs.

The length of the Park Scenic Drive is a multiplier in the product that does not have repeated digits in any row:

$$
\begin{array}{cccc}
  &   & 4 & * \\
  &   & * & * \\
\hline
  & 3 & * & * \\
  & * & 0 &   \\
\hline
* & * & * & * \\
\end{array}
$$

What is the length of the Petrified Forest National Park Scenic Drive? What relations between multipliers have you noticed? How does this information affect your result?

8. Among main attractions of the 55 sq. mile *Acadia National Park,* Maine, is the scenic Loop Road that wanders through the highest point on the Atlantic Coast, the 1527-ft high Cadillac Mountain, gorgeous Sand Beach with icy water, scaring Thunder Hole, and Bass Harbor.

How long is the Loop Road if its length in miles is determined by

$$
\frac{1}{\sqrt{49}+\sqrt{50}} + \frac{1}{\sqrt{50}+\sqrt{51}} + \frac{1}{\sqrt{51}+\sqrt{52}} + \ldots + \frac{1}{\sqrt{1153}+\sqrt{1154}}
$$
$$
+ \frac{1}{\sqrt{1154}+\sqrt{1155}} + \frac{1}{\sqrt{1155}+\sqrt{1156}}.
$$

9. *Saguaro National Monument* in Arizona, named after the large Saguaro Cactus, was founded in 1933. The Saguaro Wilderness Area was added in 1975, and *Saguaro National Park* appeared in 1994. The park

consists of two districts. The East Rincon Mountain District has a one-way spectacular driving loop, the *Cactus Forest Drive*.

A walker, jogger, runner, cyclist, and driver started the Cactus Forest Drive and left the visitor center at the same time. A driver was the first to make the loop followed by a cyclist, a runner, a jogger, and a walker. The time differences (calculated from the longest one) it took them to make the loop were inverses of the first four triangular numbers, while the values of their speed were consecutive terms of an arithmetic sequence. What is the length of the Cactus Forest Drive loop?

Use the hints below to solve the problem.

1. All values of speed are integers.
2. Jogging can be considered as running not faster than 9 mi per hour.

10. The highest natural point in Texas is Guadalupe Peak in *Guadalupe Mountains National Park*. A stainless-steel pyramid, created by American Airlines in 1958, marks its summit. The peak may be climbed on a stony trail.

    The average of two numbers that represent the length of the trail to Guadalupe Peak in miles and kilometers rounded to the nearest decimals is 5.5. The digits in these two numbers arranged in increasing or decreasing order form an arithmetic sequence. What is the trail length in miles and kilometers rounded to the nearest decimals? Are all statements of the problem needed to find the length of the trail?

    Take 1 mi = 1.6 km.

11. The longest cave system on Earth in *Mammoth Cave National Park* in Kentucky. It offers miles of trails below and above the ground. The beautiful forest surrounds cave entrances. The Temple Hill Trailhead in Mammoth Cave National Park is a staring point of several trails.

    Two tourists left the Temple Hill Trailhead at the same time. They chose the McCoy Hollow trail. The first tourist walked at 1.5 mph. In 10 minutes the third tourist left the Temple Hill Trailhead to walk the same trail. The third tourist caught up the first tourist soon and, in 20 minutes after this, he met the second tourist. It took the third tourist 2 hours and 6 minutes to cover the trail. The speed values of the three tourists are consequent terms of an arithmetic sequence. What is the length of the McCoy Hollow trail?

Answers

1. 1050 mi

2. 54 mph, 36 mph, no

3. 54 mph, 36 mph, 216 mi

4. 54 mph, 36 mph; B, E, and F; B, E, and H; C and D; etc.

5. 334.84 mi

6. 48 mph, 36 mph

7. 28 mi or 45 km

8. 27 mi

9. 8 mi

10. 4.2 mi, 6.8 km, no

11. 6.3 mi

Solutions

1. 1st way. Let x mi be the distance covered during the first day. Then $35x$ mi is the distance between Grand Canyon and Glacier National Parks and $x + (2x + 2) + 3(2x + 2) + 4 + 4(3(2x + 2) + 4) + 8 = 35x$. Then $33x + 60 = 35x$ or $x = 30$. Hence, $35x$ is $30 \cdot 35 = 1050$.

   2nd way. Let x mi be the total distance, then

$$x = \frac{1}{35}x + 2 \cdot \frac{1}{35}x + 2 + 3 \cdot \left(2 \cdot \frac{1}{35}x + 2\right) + 4 + 4 \cdot \left(3 \cdot \left(2 \cdot \frac{1}{35}x + 2\right) + 4\right)$$
$$+ 8 \Rightarrow x = 1050.$$

   The driving distance between Grand Canyon and Glacier National Parks is 1050 mi.

2. Let u and v be average speed of both cars when they left the North Rim. In 2 hours they were $2u - 2v = 36$mi apart. Since 20 minutes is 1/3 of an hour, the second car covered $\frac{4}{3}v$ mi during the third hour and $\frac{4}{3} \cdot \frac{4}{3}v$ mi during the fourth hour. The cars were 32 mi apart in 4 hours, or $4u - \left(2v + \frac{4}{3}v + \frac{4}{3} \cdot \frac{4}{3}v\right) = 32$. We obtain a system of two equations in two variables:

$$\begin{cases} 2u - 2v = 36 \\ 4u - \left(2v + \frac{4}{3}v + \frac{4}{3}\cdot\frac{4}{3}v\right) = 32 \end{cases}$$ Simplifying the second equation we

get

$$\begin{cases} u - v = 18 \\ 2u - \frac{23}{9}v = 16 \end{cases} \Rightarrow \begin{array}{l} u = 54 \\ v = 36 \end{array}.$$

It is not possible to find the driving distance between the North Rim and South Rim based on the information provided.

3. Let u and v be average speed of both cars when they left the North Rim, then $2u - 2v = 36$. The first car spent 4.5 hours on the road, 4 hours of driving with a 30-minute stop. Then the driving distance between the North Rim and South Rim is $4u$ mi. Since 20 minutes is $1/3$ of an hour, the second car covered $\frac{4}{3}v$ mi during the third hour and $\frac{4}{3}\cdot\frac{4}{3}v$ mi during the fourth hour. The second driver drove the last 30 minutes or ½ hour (two cars arrived at the South Rim at the same time) with $\frac{4}{3}\cdot\frac{4}{3}v$ mph covering distance of $\frac{1}{2}\cdot\frac{4}{3}\cdot\frac{4}{3}v$ mi. Hence, the driving distance of the second car can be presented as $2v + \frac{4}{3}v + \frac{4}{3}\cdot\frac{4}{3}v + \frac{1}{2}\cdot\frac{4}{3}\cdot\frac{4}{3}v$ mi. Thus, we obtain the system of two linear equations in two variables:

$$\begin{cases} 2u - 2v = 36 \\ 4u = 2v + \frac{4}{3}v + \frac{4}{3}\cdot\frac{4}{3}v + \frac{1}{2}\cdot\frac{4}{3}\cdot\frac{4}{3}v \end{cases} \Rightarrow \begin{array}{l} u = 54 \\ v = 36 \end{array}.$$ The distance is $54 \cdot 4 = 216$ mi.

The driving distance between the North Rim and South Rim is 216 mi.

4. Let u and v be average speed of both cars when they left the North Rim.

   1. B, E, and F $\Rightarrow$ the speed of the first car $216/4 = 54$ mph; 22 min and 30 sec $= 3/8$. Let v be the initial speed of the second car $\Rightarrow 2v + \frac{4}{3}v + \frac{4}{3}\cdot\frac{4}{3}v + \frac{3}{8}\cdot\frac{4}{3}\cdot\frac{4}{3}\cdot\frac{4}{3}v = 216 \Rightarrow v = 36$.

   2. B, E, and H $\Rightarrow$ the speed of the first car $216/4 = 54$ mph; $54 + 10 = 64 \Rightarrow \frac{4}{3}\cdot\frac{4}{3}v = 64 \Rightarrow v = 36$.

   3. C and D see the solutions to the previous problems of this section.

   4. Hint A does not provide any useful information for the problem solution.

The driving distance between the North Rim and South Rim is 216 mi.

5. $1 \text{ km} = \frac{382.28}{615.22} \text{ mi}$; $1 \text{ nautical mi} = \frac{554.285}{299.29} \text{ km} = \frac{554.285}{299.29} \cdot \frac{382.28}{615.22} \text{ mi}$;

$290.97 \text{ nautical mi} = 290.97 \cdot \frac{554.285}{299.29} \cdot \frac{382.28}{615.22} \text{ mi} = 334.84 \text{ mi}$.

6. 15 minutes is ¼ of an hour. Let u and v be the average speed of first and second cars. The first car from Chaco drove for (1 + 1 and ¼) hours before meeting the second car that left Mesa Verde. Then

$$\begin{cases} \left(1 + 1\frac{1}{4}\right)u + 1v = 144 \\ 1\frac{1}{2}u + 2v = 144 \end{cases} \Rightarrow [\text{Eq2.} -2\text{Eq}.1] \Rightarrow \begin{cases} \frac{9}{4}u + v = 144 \\ -3u = -144 \end{cases} \Rightarrow \begin{matrix} u = 48 \\ v = 36 \end{matrix}.$$

7. The solution may vary.

1. The first digit in the last row is 1 because 3 added to any one-digit number is less than 12.

2. To keep a two-digit number in the 4th row and a five-digit number in the 5th row, the first digit in the 2nd row is 2.

3. To get 0 in the 4th row, the last digit in the 1st row should be 0 or 5. Then the first row can be 40 or 45. The first option 40 does not work because 2 · 40 = 80 and 3 + 8 = 11 that lead to repeating digits in the last row.

4. The second digit in the 2nd row can be 7, 8, or 9, because only then 4 times a one-digit number plus a possible carrying digit will be 30 or higher.

5. The possible multipliers (the first and second numbers) can be (45, 27), (45, 28), (45, 29). Only (45, 28) has 3 as the first digit in the 3rd row and does not have repeated digits in the last row.

Finally, 45/28 = 1.667 is as 1 km to 1 mi.

The length of the Petrified Forest National Park Scenic Drive is given in kilometers and miles.

8. Using $a^2 - b^2 = (a - b)(a + b)$ and $\frac{1}{(a+b)} = \frac{(a-b)}{(a-b)(a+b)} = \frac{a-b}{a^2-b^2}$ (see Appendix I-AF), we get

$$\frac{1}{\sqrt{49} + \sqrt{50}} + \frac{1}{\sqrt{50} + \sqrt{51}} + \frac{1}{\sqrt{51} + \sqrt{52}} + \ldots + \frac{1}{\sqrt{1153} + \sqrt{1154}}$$

$$+ \frac{1}{\sqrt{1154} + \sqrt{1155}} + \frac{1}{\sqrt{1155} + \sqrt{1156}}$$

$$= \frac{\sqrt{49} - \sqrt{50}}{(\sqrt{49} + \sqrt{50})(\sqrt{49} - \sqrt{50})} + \frac{\sqrt{50} - \sqrt{51}}{(\sqrt{50} + \sqrt{51})(\sqrt{50} - \sqrt{51})}$$

$$+ \ldots + \frac{\sqrt{1155} - \sqrt{1156}}{(\sqrt{1155} + \sqrt{1156})(\sqrt{1155} - \sqrt{1156})} = -\frac{\sqrt{49} - \sqrt{50}}{1}$$

$$- \frac{\sqrt{50} - \sqrt{51}}{1} - \ldots - \frac{\sqrt{1154} - \sqrt{1155}}{1} - \frac{\sqrt{1155} - \sqrt{1156}}{1}$$

$$= -\sqrt{49} + \sqrt{1156} = 27.$$

The Loop Road in Acadia National Park is 27 mi.

9. The first four triangular numbers are 1, 3, 6, 10 (see Appendix I-SS). Let $s, v, d > 0$ be the loop length, the walker speed, and the difference of the arithmetic sequence correspondently, then

$$\begin{cases} \frac{s}{v} - \frac{s}{v+d} = 1 \\ \frac{s}{v+d} - \frac{s}{v+2d} = \frac{1}{3} \\ \frac{s}{v+2d} - \frac{s}{v+3d} = \frac{1}{6} \\ \frac{s}{v+3d} - \frac{s}{v+4d} = \frac{1}{10} \end{cases}$$. The first two equations lead to

$$\begin{cases} sd = v(v+d) \\ 3sd = (v+d)(v+2d) \end{cases}$$. Substituting sd from the first equation to

the second, we obtain $\frac{1}{3} = \frac{v}{v+2d} \Rightarrow d = v$.

Note: The same result appears if any other pairs of two equations are taken.

Substituting $v = d$ to the first (or any) equation we obtain $s = 2v$. Then the average speed of a walker is v, a jogger is $2v$, a runner is $3v$, a cyclist is $4v$, and a driver is $5v$.

All numbers that represent the speed are integers. A jogger's speed can reach 9 mph. On the other side, it should be even (to have an integer walker's speed) and $3v$ be greater than 9 (to have a runner's speed greater than 9). The only number that satisfies these conditions is 8. Then $v = 4$ and $s = 12$.

Remark. A triangular number is expressed (see Appendix I-SS) as $T_n = \frac{n(n+1)}{2} \Rightarrow \frac{1}{T_n} = \frac{2}{n(n+1)}$, $n = 1, 2, 3, \ldots$.

The difference in time if $s = 2v$ and $d = v$ will be always a triangular number:

$$\frac{sd}{(v + nd)(v + (n + 1)d)} = \frac{2v^2}{(v + nv)(v + (n + 1)v)} = \frac{2}{(n + 1)(n + 2)},$$

$$n = 0, 1, 2, \ldots$$

The scenic Cactus Forest Drive is 12 mi.

10. Let k and m be the length of the trail in miles and kilometers rounded to the nearest decimals, then $\frac{k+m}{2} = 5.5$, or $k + m = 2 \cdot 5.5$. Remembering that $k = 1.6m$, we get $m + 1.6m = 11$, that lead to $m = 4.2$ and $k = 6.8$.

The digits 2, 4, 6, and 8 form an arithmetic sequence. This fact is not necessary to solve the problem. Thus, this condition is extra.

The trail to the top of Guadalupe Peak is 4.2 mi or 6.8 km long.

11. Let d mph and t h be the difference of an arithmetic sequence and time the first tourist walked before the third tourist met him/her. Since the speeds of three tourists satisfy an arithmetic sequence (see Appendix I-SS), the second tourist walked with $(1.5 + d)$ mph and the third tourist with $(1.5 + 2d)$ mph. The first tourist walked $1.5t$ mi before meeting with the third tourist that left 10 minutes or $1/6$ of an hour later, but walked the same distance, that is, $1.5t = \left(t - \frac{1}{6}\right)(1.5 + 2d)$ mi. The second tourist walked 20 min or 1.3 of an hour $\left(t + \frac{1}{3}\right)(1.5 + d)$ mi before meeting with the third tourist that walked $\left(t - \frac{1}{6} + \frac{1}{3}\right)(1.5 + 2d)$ mi.

Then $\begin{cases} 1.5t = \left(t - \frac{1}{6}\right)(1.5 + 2d) \\ \left(t + \frac{1}{3}\right)(1.5 + d) = \left(t - \frac{1}{6} + \frac{1}{3}\right)(1.5 + 2d) \end{cases} \Rightarrow \begin{cases} 24dt = 4d + 3 \\ 4dt = 1 \end{cases} \Rightarrow d = .75.$

Thus, the third tourist walked $1.5 + 2 \cdot .75 = 3$ mph.

The length of the McCoy Hollow trail is $3 \cdot 2\frac{1}{10} = 6.3$ mi.

IV.2 Across the States

Highways, roads, and trails, like blood vessels in the body, go through the country connecting its states and cities.

1. The *longest* distance between two mainland points within 48 contiguous states is between Point Arena, CA and West Quoddy Head, ME.

It took Robert 53 hours to drive from Point Arena to West Quoddy Head. Robert drove a third of the road trip at 84 mi per hour, a quarter of the remaining part at 49 mi per hour, 6/7 of the remaining road at 72 mi per hour, and its last section at 42 mi per hour.

Highways bend and do not reflect a straight-line distance between cities. If it were a highway that connects these two points with a straight line, and if Robert drove with the same pattern, he would spend 9 hours and 33 minutes less on the road. What are the driving and straight distances between Point Arena and West Quoddy Head?

2. The distance between Columbus, the capital of Ohio, and Pittsburgh, Pennsylvania, is 190 mi. The distance between Pittsburgh and the nation capital Washington, D.C. is 250 mi. Let us consider a triangle CPW, where vertices C, P, W stand for Columbus, Pittsburgh, and Washington. The median from the vertex P to the side CW is $20\sqrt{13}$ mi. How far is Columbus from Washington? What is the area of the triangle CPW?

3. The distance between Youngstown, Ohio, and West Middlesex, Pennsylvania, is 16 mi. The cities form a triangle YNW with the first letter of a city as its vertex and N stands for New Castle, Pennsylvania. The $3\sqrt{21}$ mile-long bisector from the angle N divides the side WY in the ratio 7 to 9 (from the point W). What are the distances between Youngstown and New Castle and between New Castle and West Middlesex?

4. Find the distances between Youngstown, Ohio, and New Castle, Pennsylvania, and between New Castle and West Middlesex, Pennsylvania if the cities form a triangle YNW with the first letter of a city as its vertex. All sides of the triangle YNW are integers and the bisector NB from the angle N divides the side WY in the ratio 7 to 9 (from the point W). The distance between Youngstown and West Middlesex is 16 mi. Describe the triangle YNW by to its angles so that the problem has a unique solution.

   A. The triangle YNW is an acute triangle.

   B. The triangle YNW is an obtuse triangle.

   C. The triangle YNW is a right triangle.

5. Youngstown in Ohio, West Middlesex and New Castle in Pennsylvania form a triangle YNW with the first letter of a city as its vertex. The median from the vertex N to the side WY is 14 mi. The bisector from this angle divides the side WY in the ratio 7 to 9 (from the point W). The distance between Youngstown and West

Middlesex is 16 mi. What are the distances between Youngstown and New Castle, Pennsylvania and between New Castle and West Middlesex?

6. Find the driving distance between Raleigh, the capital of South Carolina, and Wilmington, the city at the shore. Use just one of the following hints:

   i. It takes 2 hours to drive from Raleigh to Wilmington.

   ii. The driving distance is shortened from ¼ to 1/3 within 11 minutes.

   iii. 11 mi shorten the driving distance from ¼ to 1/3.

   Explain why other hints are not necessary to find a solution.

7. A group of officials rushed from Kearney, NE, to a meeting in the state capital Lincoln. Knowing the estimated driving time to Lincoln, the group planned accordingly. However, because of a road repair, the group had to stop for 10 minutes and 30 seconds. Then they drove faster to arrive at the meeting on time. Find the driving distance between Kearney and Lincoln using the following hints if necessary:

   i. It usually takes 2 hours to drive from Kearney to Lincoln.

   ii. If the group continue driving with the previous speed, they would be 10 and half minutes late for the meeting.

   iii. The group drove 12 minutes before they had to stop due to a road repair.

   iv. The group drove 7 mph faster after the stop to arrive at the meeting on time.

   Explain why all hints are not needed.

8. The distance between two state capitals, the capital of Virginia, Richmond, and the capital of North Carolina, Raleigh, is 146 mi. The distance between Richmond and Wilmington, North Carolina, is 257 mi. How far are Raleigh and Wilmington, if the medians RN and WM from R and W in the triangle VRW are perpendicular? The vertices V, R, and W stand for Richmond, Raleigh, and Wilmington.

9. The distances from Fargo, North Dakota, to two towns in Minnesota, International Falls and Duluth, are the same and equal to 257 mi. The straight lines that connect these towns form a triangle FID, where F, I, and D are the first letter of the

corresponding city. The distance PQ between the end points of two bisectors DP and IQ to the sides FI and FD is 100.2 mi. What is the distance between International Falls and Duluth?

10. The distances between three major cities in Missouri, Kansas City, St. Louis, and Springfield, are 254 mi (Kansas City–St. Louis), 208 mi (Springfield–St. Louis), 174 mi (Kansas City–Springfield). These cities form a triangle with the state capital, Jefferson City, inside. Jefferson City is 133 mi away from Springfield and from St. Louis. Is Jefferson City as far from Kansas City as from Springfield and St. Louis? If not, then how far from these cities should the capital be located to have the same distance to the three major cities in Missouri?

11. The distances between three major cities in Missouri, Kansas City, St. Louis, and Springfield are 254 mi (Kansas City–St. Louis), 208 mi (Springfield–St. Louis), 174 mi (Kansas City–Springfield). These cities form a triangle. How far from these cities should the capital be located to have the same distance to these major cities?

12. The distance between Waterville, Maine, and Portsmouth, New Hampshire, can be divided by Augusta, the capital of Maine, and Portland, Maine, into three driving segments: from Waterville to Augusta, from Augusta to Portland, and from Portland to Portsmouth. The sum of the lengths of the first and third segments is related to the length of the second segment as 7 to 5. The average speed is different at these segments. Surprisingly, the driving speed on the first segment is the average of the speed values at the second and third segments. It takes 5 minutes less to drive the third segment than the second one, which is one mile longer than the third segment. It takes 1 hour and 10 minutes to drive from Waterville to Portland (the first and second segments) and 1 hour and 35 minutes to drive from Augusta to Portsmouth (the second and third segments). What are the driving time and total distance between Waterville and Portsmouth?

13. Durango, Colorado, is a gate to the marvelous *Mesa Verde National Park* famous for its well-preserved Ancestral Puebloan cliff dwellings.

The distances between Durango and Santa Fe, the capital of New Mexico, and between Durango and Albuquerque, the largest city in New Mexico, are approximately the same. The median and height to the side Durango–Santa Fe in the triangle with the vertices in Durango, Albuquerque, and Santa Fe are m_a and h_a miles. Express

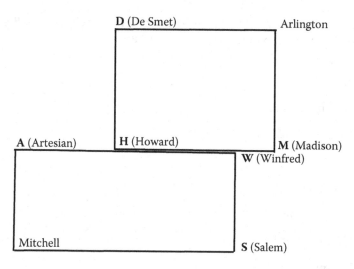

FIGURE IV.1

the distance between Santa Fe and Albuquerque and between Santa Fe and Durango in terms of m_a, h_a. How far are these cities from each other if $m_a = 113.657$ and $h_a = 57.455$?

14. Main roads in South Dakota are like straight lines and may be viewed as a combination of different rectangles connected to each other, see Figure IV.1. Mitchell, Salem (S), Winfred (W), and Artesian (A) form one rectangle. Howard (H), Madison (M), Arlington, and De Smet (D) form a smaller rectangle. The difference in the areas of the two rectangles is 220 sq. mi. Madison, Winfred, Howard, and Artesian are located on the same road (line). The distance between Madison and Winfred is 4 mi more than between Winfred and Howard. The distance between Howard and Artesian is 1 mile more than between Howard and De Smet. The distances between Salem and Winfred and between Howard and Madison are the same. The driving distance on the outer loop (without driving from Winfred to Howard) is 202 mi. What are the distances between the towns?

15. The distance between Topeka, the capital of Kansas, and Fairview, a small city in Kansas is 95 km. Two bikers left these cities at the same time and biked toward each other. The biker that left Topeka biked 1 km/h faster than the one who left Fairview. They met in the number of hours that is a half of the speed value of the Topeka

biker. In how many hours did the bikers meet? What is the distance in miles between Topeka and Fairview?

Answers

1. 3528 mi, 2892 mi

2. 420 mi, 13, 630 sq. mi

3. 18 mi, 14 mi

4. 18 mi, 14 mi, A

5. 18 mi, 14 mi

6. 132 mi, Hint iii

7. 130 mi

8. 132.19 mi

9. 164 mi

10. 121.77 mi, 128 mi

11. 128 mi

12. 1 h 55 min, 132 mi

13. $AD = DS = \frac{2}{3}(\sqrt{m_a^2 - h_a^2} + \sqrt{4m_a^2 - h_a^2})$;
    $AS^2 = (a - \sqrt{m_a^2 - h_a^2})^2 + h_a^2$, 58 mi, 212 mi, 212 mi

14. 9 mi, 22 mi, 28 mi, 38 mi

15. 5 h, 59.03 mi

Solutions

1. Let x be the driving distance. Robert drove $1/3x$ mi at 84 mph, ¼ of the remaining $(1 - 1/3)x = 2/3x$ mi, that is, $1/6x$ mi at 49 mph, $6/7$ of $(1 - 1/3 - 1/6)x$ mi or $3/7x$ mi at 72 mph, and remaining $(1 - 1/3 - 1/6 - 3/7)x = 1/14x$ mi at 42 mph. Thus,

$$\frac{\frac{1}{3}x}{84} + \frac{\frac{1}{6}x}{49} + \frac{\frac{3}{7}x}{72} + \frac{\frac{1}{14}x}{42} = 53 \Rightarrow \frac{14 + 12 + 21 + 6}{2 \cdot 6 \cdot 6 \cdot 7 \cdot 7}x = 53 \Rightarrow x = 3528.$$

53 hours without 9 hours and 33 minutes is 43.45 hours. Then the straight distance is $\frac{3528 \cdot 43.45}{53} \approx 2892$.

The longest straight distance between two US mainland points, Point Arena and West Quoddy Head is 2892 mi, while the driving distance is 3528 mi.

2. The relation among the median m_a to the side a and the sides a, b, and c (see Appendix I-T) is $4m_a^2 = 2b^2 + 2c^2 - a^2$. Then, $CW^2 = 2PC^2 + 2PW^2 - 4PM^2 = 176400$, or $CW = 420$ mi and $A = \sqrt{p(p-a)(p-b)(p-c)} \approx 13{,}630$ sq. mi, where $p = 430$ mi is a semi perimeter of a triangle.

Columbus is 420 mi far from Washington.

3. Let B be the end point of the bisector from the angle N to the side WY. Combining $WB/BY = 7/9$ and $WB + BY = 16$ we get $WB = 7$ and $BY = 9$.

A bisector of a triangle divides the opposite side of a triangle into segments proportional to the adjacent sides, that is $NW/NY = 7/9$ (see Appendix I-T). Hence, the sides can be presented as $NW = 7n$, $NY = 9n$, where n is any real number. It is obvious that $n > 1$, otherwise it is impossible to have a triangle. Then the formula for the bisector $l_c^2 = ab - a_1 b_1$ gives $189 = 7n \cdot 9n - 7 \cdot 9 \Rightarrow 189 = 7 \cdot 9(n^2 - 1) \Rightarrow n = 2$ that leads to the result.

The distance between Youngstown and New Castle is 18 mi and between New Castle and West Middlesex is 14 mi.

4. From the problem follows that $WB/BY = 7/9$ and $WB + BY = 16$. Then, $WB = 7$, $BY = 9$. A bisector of a triangle divides the opposite side of triangle into segments proportional to the adjacent sides (see Appendix I-T), that is, $NW/NY = 7/9$. Thus, the sides can be presented as $NW = 7n$, $NY = 9n$, where n is an integer. It is obvious that $n > 1$, otherwise it is impossible to have a triangle, because the third side will be greater than the sum of the other two.

Let us consider Hint C. The largest side NY is a hypotenuse and, according to the Pythagoras theorem, $16^2 = NY^2 - NW^2 = 32n^2$, which does not have an integer solution n. Hint C is not applicable. Now let us draw the height NH in the triangle YNW (see Figure IV.2), and consider two right triangles $\triangle YHN$ and $\triangle NHW$ then $NH^2 = NW^2 - WH^2$ and $NH^2 = NY^2 - YH^2 \Rightarrow NW^2 - WH^2 = NY^2 - YH^2 \Rightarrow 32n^2 = YH^2 - WH^2 \Rightarrow 32n^2 = 16(16 \pm 2WH) \Rightarrow n^2 = 8 \pm WH$, where "−" is for the acute triangle YNW and "+" is for the obtuse one.

The equation $n^2 = 8 - WH$ has the integer solution $n = 2 > 1$ if $WH = 4$. Then the triangle YNW is an acute triangle (Hint A).

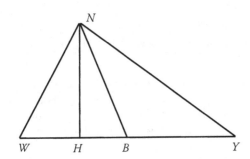 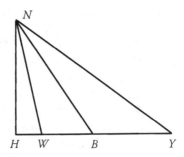

FIGURE IV.2

The equation $n^2 = 8 + WH$ has integer solutions $n = 3 > 1$ at $WH = 1$, $n = 4 > 1$ at $WH = 8$, and so on. Thus, the solution is not unique when the triangle YNW is obtuse. So, Hint B does not work.

The distance between Youngstown and New Castle is 18 mi and between New Castle and West Middlesex is 14 mi.

5. A bisector of a triangle divides the opposite side of triangle into segments proportional to the adjacent sides, that is $NW/NY = 7/9$. Thus, the sides can be presented as $NW = 7n$; $NY = 9n$, where n is a real number. Applying the formula for the median (see Appendix I-T) $4m_a^2 = 2b^2 + 2c^2 - a^2$, we get $4{\cdot}14^2 = 2{\cdot}(9n)^2 + 2{\cdot}(7n)^2 - 16^2 \Rightarrow n = 2$. Then $NW = 14$ and $NY = 18$.

The distance between Youngstown and New Castle is 18 mi and between New Castle and West Middlesex is 14 mi.

6. The distance part $\frac{1}{3} - \frac{1}{4} = \frac{1}{12}$ represents 11 mi by Hint iii. Therefore, the total distance is $11 \cdot 12 = 132$.

The driving distance between Raleigh and Wilmington is 132 mi. Hints i and ii do not provide any connection to distance.

7. Let u mph be the driving speed, then $2u$ mi is the distance between Kearney and Lincoln by Hint i. They drive after the accident at $u + 7$ by Hint iv. Using Hint iii and transforming 12 minutes to 1/5 of an hour and 10 minutes 30 seconds to 7/40 of an hour (then $2\,h - 1/5\,h - 7/40\,h = 13/8\,h$), we can write $\frac{1}{5}u + \frac{13}{8}(u + 7) = 2u \Rightarrow \frac{7}{40}u = \frac{91}{8} \Rightarrow u = 65$.

The driving distance between Kearney and Lincoln is 130 mi.

8. Let x mi be RW, the distance between Raleigh and Wilmington, and p and q denote ML and LN, where L is the point of intersections of the medians RN and WM, as shown in Figure IV.3. Because

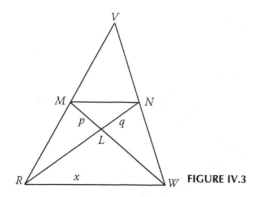

FIGURE IV.3

$VM = \frac{1}{2}RV$, $VN = \frac{1}{2}VW$, $\angle V$ is common $\Rightarrow \Delta RVW$ and ΔMVN are similar. Thus, MN and RW are parallel, and $MN = RW/2$. Then, ΔMLN and ΔWLR are also similar and $ML = \frac{1}{2}LW$, $LN = \frac{1}{2}RL \Rightarrow LW = 2p$, $RL = 2q$. Because MW and RN are medians that divide the sides RV and VW in halves, and perpendicular, then, from three right triangles ΔRLW, ΔMLR, ΔLNW, we get:

$$\begin{cases} x^2 = (2p)^2 + (2q)^2 \\ \left(\frac{1}{2}\cdot 257\right)^2 = (2p)^2 + q^2 \\ \left(\frac{1}{2}\cdot 146\right)^2 = p^2 + (2q)^2 \end{cases} \Rightarrow \begin{cases} x^2 = 4(p^2 + q^2) \\ \frac{1}{4}(257^2 + 146^2) = 5(p^2 + q^2) \end{cases} \Rightarrow x^2 = \frac{257^2 + 146^2}{5}$$

$$\Rightarrow x = 132.19.$$

The distance between Raleigh and Wilmington is 132.19 mi.

9. Let x be the distance ID between International Falls and Duluth, and P and Q be the end points of the bisectors to FI and FD (see Figure IV.4). Because IQ is a bisector of a triangle, then, by the angle bisector theorem, it divides the side FD in proportion $\frac{IF}{ID} = \frac{FQ}{QD} \Rightarrow \frac{257}{x} = \frac{p}{257 - p}$, where p stands for FQ. Because the ΔIFD is isosceles, PQ and ID are parallel (IQ and DP are equal in an isolated triangle), and ΔPFQ and ΔIFD are similar, then $\frac{PQ}{ID} = \frac{FQ}{FD} \Rightarrow \frac{100.2}{x} = \frac{p}{257}$. Combining two proportions, we get

$$\begin{cases} \frac{p}{257 - p} = \frac{257}{x} \\ \frac{100.2}{x} = \frac{p}{257} \end{cases} \Rightarrow \begin{cases} \frac{257 - p}{p} = \frac{x}{257} \\ \frac{100.2}{x} = \frac{p}{257} \end{cases} \Rightarrow \begin{cases} \frac{257}{p} = \frac{x + 257}{257} \\ \frac{100.2}{x} = \frac{p}{257} \end{cases} \Rightarrow \frac{x}{100.2} = \frac{x + 257}{257} \Rightarrow x \approx 164.$$

The distance between International Falls and Duluth is approximately 164 mi.

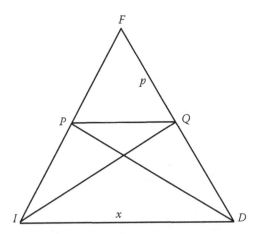

FIGURE IV.4

10. Let us draw a triangle and use the first letters of the cities for the vertices of the triangle, then $KS = 174$, $KL = 254$, $LS = 208$, $JS = JL = 133$ (see Figure IV.5). Applying the theorem of cosines three times $\cos \angle JSL = \frac{SL^2 + SJ^2 - JL^2}{2SL \cdot SJ} = \frac{SL}{2SJ} = 0.78195 \Rightarrow \angle JSL = 38.56°$ and $\cos \angle KSL = \frac{KS^2 + SL^2 - KL^2}{2KS \cdot SL} = 0.124668 \Rightarrow \angle KSL = 82.838° \Rightarrow \angle KSJ = 44.278° \Rightarrow KJ^2 = KS^2 + SJ^2 - 2KS \cdot SJ \cos \angle KSL \Rightarrow KJ = 121.77$. The distance from Jefferson City to Kansas City is 121.77 mi.

The center of the circumscribed circle is at the same distance from all vertices of a triangle and its radius is the shortest distance to the vertices. From the theorem of sines, $2R = \frac{KL}{\sin \angle KSL} \Rightarrow R = 127.999$. The distance from the capital to the three major cities would be 128 mi.

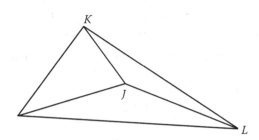

FIGURE IV.5

11. The center of the circumscribed circle is of the same distance from all vertices of a triangle. Thus, the radius of the circle is the shortest distance to the vertices. Combining two formulas $A = \sqrt{p(p-a)(p-b)(p-c)}$, $R = \frac{abc}{4A} \Rightarrow R = \frac{abc}{4\sqrt{p(p-a)(p-b)(p-c)}} \Rightarrow$ $R = \frac{9,192,768}{4 \cdot 192\sqrt{8745}} \approx 128$, where the semi perimeter $p = 318$.

The distance from the capital should have been 128 mi

12. Let us divide the problem into two parts: finding the driving time and then the total distance. Rewriting in hours: 5 minutes = 1/12 of an hour, 1 hour and 10 minutes = 7/6, 1 hour and 35 minutes = 19/12, denoting the lengths of the segments as l_1, l_2, l_3, and following the problem, we can prepare the table to have equidistance from the three major cities.

Driving Distance in Miles	Time in Hours
$l_1 + l_2$	7/6
$l_2 + l_3$	19/12
$l_2 - l_3$	1/12

from which follows that it took 4/12 = 1/3 hours to drive the first segment of the road, 10/12 = 5/6 h to drive the second, and 9/12 = 3/4 h – the third. Thus, the total driving time is 23/12 hours = 1 and 55/60 hours = 1 hour and 55 minutes.

Considering that the sum of the lengths of the first and third segments is related to the length of the second segment as 7 to 5 and that the second segment of the road is one mile longer than the third one, we use $1 + 2l_2/5$, l_2, $l_2 - 1$ to present the lengths of the three segments. Because the driving speed on the first segment is the average of the speeds on the second and third segments, we can write $(1 + 2l_2/5)/(1/3) = [l_2/(5/6) + (l_2 - 1)/(3/4)]/2$, from which follows that $l_2 = 55$, $l_1 = 23$, and $l_3 = 54$.

Thus, the driving time is 1 hours and 55 minutes and the total distance between Waterville and Portsmouth is 132 mi.

13. Let us use the first letters of a city for the vertices of the formed triangle, $AM = m_a$ for the median, and $AH = h_a$ for the height, see Figure IV.6. $\triangle AHD$, $\triangle AHM$, and $\triangle AHS$ are right triangles. Following the Pythagorean theorem, from $\triangle AHM \Rightarrow MH^2 = AM^2 - AH^2$ and from the $AD^2 = AH^2 + DH^2 = (AD/2 + MH)^2 + AH^2$. Rewriting these formulas in terms of a, where $2a$ is the side $AD = DS$, and $DM = MS = a$, we get $MH = \sqrt{m_a^2 - h_a^2}$. From $\triangle AHD$ follows

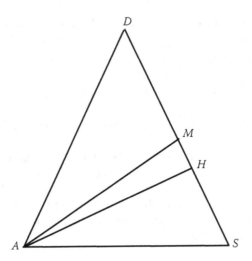

FIGURE IV.6

$4a^2 - (a + \sqrt{m_a^2 - h_a^2})^2 = h_a^2 \Rightarrow 3a^2 - 2a\sqrt{m_a^2 - h_a^2} - m_a^2 = 0 \Rightarrow$

$a = \dfrac{\sqrt{m_a^2 - h_a^2} \pm \sqrt{4m_a^2 - h_a^2}}{3}$. Only $AD = DS = \frac{2}{3}(\sqrt{m_a^2 - h_a^2} + \sqrt{4m_a^2 - h_a^2})$

satisfies our problem because $\frac{2}{3}(\sqrt{m_a^2 - h_a^2} - \sqrt{4m_a^2 - h_a^2})$ is negative and cannot be a side (see Appendix I-T). From $\triangle AHS$ follows $AS^2 = (a - \sqrt{m_a^2 - h_a^2})^2 + h_a^2$.

Substituting $m_a = 113.657$ and $h_a = 57.455$ to the formulas, we get that the distance between Santa Fe and Albuquerque is 58 mi, and between Durango and either Santa Fe or Albuquerque is 212 mi.

The problem can be solved using different ways, for instance, using the relation between the median and the sides is $4m_a^2 = 2b^2 + 2c^2 - a^2$ that in our case turns to be $4 \cdot AM^2 = DS^2 + 2AS^2$.

The distance between Santa Fe and Albuquerque is 58 mi, between Durango and either Santa Fe or Albuquerque is 212 mi.

14. $A_2 - A_1 = AW \cdot SW - HM \cdot HD = 220$.   Since $AW = HD + 1 + (HM - 4)/2$, and $SW = HM$, then $HM((HD + 1) + (HM - 4)/2 - HD) = HM(HM - 2)/2 = 220 \Rightarrow HM^2 - 2HM - 440 = 0 \Rightarrow HM = 1 \pm \sqrt{1 + 440} \Rightarrow HM = 22 \Rightarrow HW = 9, WM = 13$. Next, the outer loop $\Rightarrow 13 + 2MA + 22 + (MA + 1) + 2 \cdot 22 + (MA + 10) = 202 \Rightarrow MA = 28$.

The distances between Salem and Winfred, Arlington and De Smet, Mitchell and Artesian, and Howard and Madison are 22 mi,

Topecka   Fairview **FIGURE IV.7**

between Mitchell and Salem and between Winfred and Artesian
are 38 mi, between De Smet and Howard and Arlington and
Madison are 28 mi, and between Winfred and Howard is 9 mi.

15. Let v km/h be the speed of the Topeka biker, then $(v - 1)$ km/h is
the speed of the Fairview biker, see Figure IV.7. They met in $v/2$
hours. Hence, $v \cdot \frac{v}{2} + (v - 1) \cdot \frac{v}{2} = 95 \Rightarrow 2v^2 - v - 190 = 0 \Rightarrow v = \frac{1 \pm 39}{4} \Rightarrow v = 10$. The bikers met in 5 hours. The distance in miles is
$95 \cdot 0.621371 = 59.03$.

IV.3 The US Highways

The USA has a well-developed network of highways. They are recognized by their
lengths, number of lines, and speed limits. Their development and names are a
subject of numerous stories.

1. The *Red Rock Scenic Byway* winds through Sedona's Red Rock
   Country and is often called a *museum without walls*.

   The Red Rock Scenic Byway is only 7.5 mi from Sedona to I-17.
   Two hikers walked along the byway. They left Sedona at the same
   time. The first hiker took a 15-minute break after every 1.5-mi walk
   to explore amazing red rocks. Each hour the second hiker walked a
   half of a mile less than the first hiker and stopped just once for a
   half of an hour. Both hikers arrived at I-17 at the same time. How
   fast do they walk?

2. The *Dinosaur Diamond Prehistoric Highway* connects Grand Junction,
   Colorado, and Moab, Utah. It is one of the best places in the world
   to learn about dinosaurs.

   Bones and footprints of dinosaurs are visible along the 480-mi long
   Dinosaur Diamond Prehistoric Highway. A couple was excited
   about this journey and completed its travel at 48 mph, even though
   the speed was different on three unequal segments of their drive. It
   took the couple twice the time to drive the first half of the road (the
   first segment) than its last third of the road (the third segment). The
   average speed on the third segment was 16 mph more than on the

first segment. What was the average speed on the second segment of the journey?

3. Although the entire stretch of 25 mi of the US Route 550 in Colorado is called the *Million Dollar Highway*, its most famous part is only 12 mi from Ouray to the summit of Red Mountain Pass. There are several legends about its name. It is said that it cost a million dollars a mile to build it in the 1920s, or that its fill dirt contains a million dollars in gold ore, or that its roadbed was paved with low-grade gold ore.

Two cyclists biked along the 12-mi Million Dollar Highway. They left the starting point at the same time. The first cyclist biked 1.2 mph faster than the second one and arrived at the end point 30 minutes before the second cyclist. How fast was the second cyclist biking?

4. The scenic *Million Dollar Highway* goes through three mountain passes, the Coal Bank Pass, Molas Pass, and Red Mountain Pass (arranged in an increasing order of their elevations). Their average elevation is 10,876 ft, and the median is 10,970 ft. The Molas Pass is just 48 ft lower than the Red Mountain Pass. What is the elevation of each pass?

5. Some roads in the Isle of Man, Nepal, and India are without any official speed limit. Drivers can go as fast as they want or can.

How many *without speed limit* highways or their segments are in the US if this number is equal to

$$-\frac{3}{4} + \frac{1}{3} + \frac{1}{8} + \frac{1}{15} + \frac{1}{24} + \frac{1}{35} + \frac{1}{48} + \ldots?$$

Name these US roads.

6. Built in 1959, the US Interstate 10 is the *southernmost* major highway in the American Interstate Highway System. It connects the Pacific and Atlantic Oceans and runs from State Route 1 in Santa Monica, CA to I-95 in Jacksonville, FL. The section of I-10 from Houston to the city of Katy in Texas, known as the Katy Freeway, is the *widest highway in the world*.

How many lanes does the Katy Freeway have if this number can be presented by three consecutive odd numbers in two different ways: as the sum of squares of the third and first numbers or as the difference of cubes of the second and first numbers?

7. A segment of I-10 in Southern Texas has the speed limit that makes driving special.

   A van started this I-10 segment at the permitted speed (as posted in the road speed sign). In 48 minutes, a sport car left the same point but drove 10 mph faster than the van. The sport car was 4 mi from the end of the I-10 segment when the van reached its end. If the two cars left two endpoints of this I-10 segment at the same time driving toward each other, they would meet in 3.2 hours. How long is this segment of I-10? What is the legal speed there?

8. The *third-longest major east–west* Interstate Highway in the USA, I-40, is just 98.93 mi longer than the *southernmost* transcontinental highway I-10. Both highways pass California, Arizona, New Mexico, and Texas (let us call them "common" states). Eight times the length of I-40 in these states is 116.9 mi longer than five times the length of I-10 in those states.

   The rest of I-40 goes through Oklahoma, Arkansas, Tennessee, and North Carolina, and I-10 goes through Louisiana, Mississippi, Alabama, and Florida (let us call these states "not common" states). Twice the length of I-10 in these states is 65.71 mi longer than the length of I-40 in those states.

   Furthermore, twice the difference between the lengths of I-40 in "not common" states and in "common" states is the same as the difference between the lengths of I-10 in "common" states and in "not common" states without 40.26 mi. How long are I-40 and I-10?

9. Interstate 5 (I-5) is the only highway that runs from one US border with Canada to another US border with Mexico. It runs along the US West Coast parallel to the US Highway 101 and the Pacific Ocean through Washington, Oregon, and California.

   The difference of the lengths of I-5 in Oregon and Washington is 30 mi, which is 46 times smaller than its overall total length. The sum of the two lengths is 100 mi less than the half of the total length of I-5. How long is I-5? How long is I-5 in California? How long is I-5 in Oregon and Washington?

10. The main highway on the US East Coast, Interstate 95 (I-95), is running parallel to the Atlantic Ocean from Florida to Maine. It passes through New York, Philadelphia, and Washington, D.C.

    If Sam drove 5.5 mph faster than usually, then it would take him 1 hour and 40 minutes less to drive the entire I-95. However, if he drove 22 mph slower, than it would take him 10 hours more to drive the I-95 from one end to another. What is the length of I-95?

11.–15. *The longest Interstate Highway, I-90, is the northernmost coast-to-coast interstate that connects Seattle, WA and Boston, MA. I-90 passes through the Continental Divide over Homestake Pass (Montana), which is its highest point, and the Powder River Basin (Wyoming) as well as the second and fifth longest floating bridges in the world. It also crosses the Lacy V. Murrow Memorial and the Homer M. Hadley Memorial Bridges (Washington), the tunnel under the Mount Baker Ridge (Washington) and Wallace (Idaho) listed in the National Register of Historic Places. I-10 is tolled along the Jane Addams Memorial Tollway (Illinois), Indiana Toll Road, New York State Thruway, and the Pennsylvania and Massachusetts turnpikes. There are lots of scenic and remarkable spots along I-90.*

11. When crossing Lake Washington from Seattle to Mercer Island, I-90 passes the second and fifth *longest floating bridges in the world*, the Lacy V. Murrow Memorial Bridge and the Homer M. Hadley Memorial Bridge. The Evergreen Point floating bridge is the *longest bridge* in the world, while the Hood Canal Bridge or the William A. Bugge Bridge is the third longest floating bridge and the *longest floating bridge in the world* located in a saltwater tidal basin.

    The length difference between the second and fifth longest floating bridges is just 248 m, but the length of the third longest bridge is 72 m more than twice the difference between the first longest bridge and the sum of the second and fifth ones.

    Seven times the difference between the second and third longest bridges is 8 m more than the difference between the third and fifth longest bridges. Twice the length of the second longest floating bridge is 680 m more than the difference between three times the length of the fifth longest floating bridge and the length of the second longest floating bridge.

    What are the lengths of the of longest floating bridges in the world, the Evergreen Point floating bridge (the first), the Lacy V. Murrow Memorial Bridge (second), The Hood Canal Bridge (third), and the Homer M. Hadley Memorial Bridge (fifth)?

    Does the problem as stated have a unique solution? If not, use the statement that the sum of lengths of the Lacy V. Murrow Memorial Bridge and The Hood Canal Bridge is 4008 m.

12. One third of Indiana Toll Road is 13 mi more than its one quarter, which is 4.5 mi longer than one quarter of the Massachusetts turnpike. How long are the Indiana Toll Road and Massachusetts turnpike?

13. The length of the New York State Thruway in miles is a three-digit number with different digits. Its third digit (ones) is the product of two numbers. The first and the second digits are these two numbers squared. How long is the New York State Thruway if the first digit of its length is the smallest?

14. The length of the Jane Addams Memorial Tollway in Illinois is a three-digit number presented as the product of two consequent numbers, raised to the same power as the base itself (for instance, 5^5). How long is the Jane Addams Memorial Tollway?

15. The length of the Pennsylvania turnpike is 360 mi. A driver drove fast and got a ticket at the end of the turnpike. If he/she drove just 8 mph slower, which is still 2 mph above the posted speed, the driver would drive only 30 minutes longer but would not get a speeding ticket. How fast is the driver drive when he/she got a ticket? What is the posted speed along the Pennsylvania turnpike?

16. It usually takes Olga 5 days to drive I-90 from Seattle to Boston to see her sister Vicki. It takes Vicki 7 days to drive from Boston to Seattle to visit Olga. How long is I-90 if Olga covers 177.2 mi per day more than Vicki?

17. It usually takes Olga 5 days to drive from Seattle and Boston to see her siter Vicki. It takes Vicki 7 days to drive from Boston to Seattle to visit Olga. One day the girls decided to meet somewhere at I-90. Because of an urgent job assignment, Vicki left two days after Olga. In how many days after Vicky left her house did the sisters meet?

Answers

1. 3 mph, 2.5 mph

2. 32 mph

3. 4.8 mph

4. 10,640 ft, 10,970 ft, 11,018 ft

5. 0

6. 26

7. 544 mi, 80 mph

8. 2559.25 mi, 2460.32 mi

9. 1380 mi, 790 mi, no

10. 1925 mi

11. No, 4750 m, 2020 m, 1988 m, 1772 m

12. 156 mi, 138 mi

13. 496 mi

14. 108 mi

15. 80 mph, 70 mph

16. 3101 mi

17. 1.75

Solutions

1. Let u mph be the speed of the first hiker, then $(u - 0.5)$ mph is the speed of the second hiker. The first hiker took four breaks for 15 min = 1/4 h each, while the second hiker took one break for 30 min = 1/2 h. Then

$$\frac{7.5}{u} + 4 \cdot \frac{1}{4} = \frac{7.5}{u - 0.5} + \frac{1}{2} \Rightarrow u^2 - 0.5u - 7.5 = 0 \Rightarrow u = 3, -2.5.$$

Only the first solution $u = 3$ mph satisfies the problem.

The first hiker walked at 3 mph, the second hiker walked at 2.5 mph.

2. The couple completed their trip in 480/48 = 10 hours. The first segment of the journey was ½ · 480 = 240 mi, the last one was 1/3 · 480 = 160 mi. Hence, the middle segment was 480 − 240 − 160 = 80 mi. Let u mph be the average speed on the first half of the journey (the first segment), then $\frac{240}{u} = 2 \cdot \frac{160}{u + 16} \Rightarrow u = 48$ mph is the speed on this segment. It took the couple $\frac{240}{48} + \frac{160}{48 + 16} = 7.5$ hours to cover the first half and last third of the road (the first and third segments of the journey) that left 10 − 7.5 = 2.5 hours for the second segment.

The speed at the middle segment was $\frac{80}{2.5} = 32$ mph.

3. Let x mph be the speed of the second cyclist, then $\frac{12}{x} = \frac{12}{x + 1.2} + \frac{1}{2} \Rightarrow 28.8 = x^2 + 1.2x \Rightarrow x = 4.8$ or $x = -6$. The solution $x = -6$

to the quadratic equation does not satisfy applied meaning of the problem.

The second cyclist biked with 4.8 mph.

4. The median 10,970 ft is the elevation of Molas Pass. Then adding 48 to 10,970 gives 11,018 ft, which is the elevation of Red Mountain Pass. Finally, $10,876 \cdot 3 - 10,970 - 11,018 = 10,640$ ft provides the elevation of Coal Bank Pass.

5. $\frac{1}{3} + \frac{1}{8} + \frac{1}{15} + \frac{1}{24} + \frac{1}{35} + \frac{1}{48} + \ldots = \frac{1}{4-1} + \frac{1}{9-1} + \frac{1}{16-1} + \ldots + \frac{1}{n^2-1} + \ldots$

Since $\frac{1}{n^2-1} = \frac{1}{2}\left(\frac{1}{n-1} - \frac{1}{n+1}\right)$, then $\frac{1}{4-1} + \frac{1}{9-1} + \frac{1}{16-1} + \ldots + \frac{1}{n^2-1} + \ldots$ can

be rewritten as $\frac{1}{2}\left(\frac{1}{1} - \frac{1}{3} + \frac{1}{2} - \frac{1}{4} + \frac{1}{3} - \frac{1}{5} + \frac{1}{4} - \frac{1}{6} + \ldots + \frac{1}{n-1} - \frac{1}{n+1} + \ldots\right)$

$= \frac{3}{4}$, and $-\frac{3}{4} + \frac{1}{3} + \frac{1}{8} + \frac{1}{15} + \frac{1}{24} + \frac{1}{35} + \frac{1}{48} + \ldots = -\frac{3}{4} + \frac{3}{4} = 0.$

All U.S. highways have a speed limit.

6. Let a be the second odd number, then $a - 2$ and $a + 2$ are the first and third numbers. Next, $(a - 2)^2 + (a + 2)^2 = a^3 - (a - 2)^3 \Rightarrow a^2 - 4a + 4 + a^2 + 4a + 4 = a^3 - a^3 + 6a^2 - 12a + 8$. From $4a(a - 3)$ follows $a = 0$, $a = 3$. Only $a = 3$ works. Then the three consecutive odd numbers are 1, 3, 5 and $1^2 + 5^2 = 3^3 - 1^3 = 26.$

The Katy Freeway has 26 lanes.

7. Convert 48 minutes = 0.8 hour. Let v mph and t h be the speed of the van (and the allowed speed) and the time it takes for the car to drive this segment, then $\begin{cases} tv = (t - 0.8)(v + 10) + 4 \\ 3.2v + 3.2(v + 10) = vt \end{cases} \Rightarrow$

$\begin{cases} 0 = 10t - 0.8v - 4 \\ t = 6.4 + \frac{32}{v} \end{cases} \Rightarrow \begin{cases} v^2 - 75v - 400 = 0 \\ t = 6.4 + \frac{32}{v} \end{cases} \Rightarrow v = 80$, or $v = -5$

($v = -5$ does not work). Therefore, $t = 6.8$, and $s = 544$ mi.

The official speed for the 544-mi segment of I-10 is 80 mph.

8. Let x mi be the length of I-40 in "common" states, y mi be its length in "not common" states, u mi be the length of I-10 in "common" states, and v be its length in "not common" states, that is

Interstate	"Common" States	"Not Common" States
I-10	u mi	v mi
I-40	x mi	y mi

Then
$$\begin{cases} x + y = u + v + 98.93 \\ 8x - 116.9 = 5u \\ 2v - 65.81 = y \\ 2(y - x) = (u - v) - 40.26 \end{cases} \Rightarrow \begin{cases} x + y - u - v = 98.93 \\ 8x - 5u = 116.9 \\ -y + 2v = 65.81 \\ 2x - 2y + u - v = 40.26 \end{cases}.$$

Finding $u = 1.6x - 23.38$ and $y = -65.81 + 2v$ from the second and third equations and substituting them to the first and fourth equations, we obtain the system of two equations in two variables:
$$\begin{cases} x - 65.81 + 2v - 1.6x + 23.38 - v = 98.93 \\ 2x + 131.62 - 4v + 1.6x - 23.38 - v = 40.26 \end{cases} \text{ or } \begin{cases} -0.6x + v = 141.36 \\ 3.6x - 5v = -67.98' \end{cases}$$
from which follows $x = 1064.7$ and $v = 780.18$. Substituting these values back to $u = 1.6x - 23.38$ and $y = -65.81 + 2v$ gives us $y = 1494.55$ and $u = 1494.55$.

The system of four linear equations in four variables can be also solved by any of three methods discussed in detail in Problem I-1.13.
The length of I-40 is 2559.25 mi and the length of I-10 is 2460.32 mi.

9. The length of I-5 is $46 \cdot 30 = 1380$ mi. Its half with 100 mi gives the length of the I-5 in California of 790 mi. Thus, the lengths of I-5 in Oregon and Washington are not needed to solve the problem.

The I-5 in California is 790 mi.

10. Let x mi be the length of I-95 and y mph be the Sam's usual speed. Transforming 1 h 40 min to $1^2/_3$ h, we get

$$\begin{cases} \frac{x}{y} - \frac{x}{y+5.5} = 1\frac{2}{3} \\ \frac{x}{y-22} - \frac{x}{y} = 10 \end{cases} \Rightarrow \begin{cases} xy + 5.5x - xy = \frac{5}{3}y^2 + 5.5 \cdot \frac{5}{3}y \\ xy - xy + 22x = 10y^2 - 220y \end{cases} \Rightarrow \begin{cases} 11x = \frac{10}{3}y^2 + \frac{55}{3}y \\ 11x = 5y^2 - 110y \end{cases}.$$

Equating two equations, we obtain the quadratic equation $\frac{5}{3}y^2 - \frac{385}{3}y = 0$ that has two solutions $y = 77$ and $y = 0$. The second solution $y = 0$ does not satisfy the applied meaning of the problem. Then, $y = 77$ and $x = 1925$.
The I-95 is 1925 mi long.

11. Let l_1, l_2, l_3, l_5 be the lengths of the Evergreen Point floating bridge, the Lacy V. Murrow Memorial Bridge, The Hood Canal Bridge, and the Homer M. Hadley Memorial Bridge, then

$$\begin{cases} l_2 - l_5 = 248 \\ 2(l_1 - l_2 - l_5) = l_3 - 72 \\ 7(l_2 - l_3) - 8 = l_3 - l_5 \\ 2l_3 - 680 = 3l_5 - l_2 \end{cases} \Leftrightarrow \begin{cases} l_2 - l_5 = 248 \\ 2l_1 - 2l_2 - l_3 - 2l_5 = -72 \\ 7l_2 - 8l_3 + l_5 = 8 \\ l_2 + 2l_3 - 3l_5 = 680 \end{cases}.$$

Combining the first, third, and fourth equations and illuminating l_5, we obtain a system of two linear equations in two variables:

$$\begin{cases} 8l_2 - 8l_3 = 256 \\ -2l_2 + 2l_3 = -64 \end{cases} \text{ or } \begin{cases} l_2 - l_3 = 32 \\ l_2 - l_3 = 32 \end{cases}.$$

The two equations are the same. It means that the first, third, and fourth equations are linearly dependent and the original system of four equations does not have a unique solution, but rather infinitely many solutions in this case. Using the hint provided,
$$\begin{cases} l_2 - l_3 = 32 \\ l_2 + l_3 = 4008 \end{cases},$$ we obtain $l_1 = 2020$ and $l_2 = 1988$. Substituting these values to the third equation and solving it for l_5 we get $l_5 = 1772$. Finally, from the first equation follows $l_1 = 4750$.

The Evergreen Point floating bridge is 4750 m (15,580 ft), the Lacy V. Murrow Memorial Bridge is 2020 m (6620 ft), The Hood Canal Bridge is 1988 m (6521 ft), and the Homer M. Hadley Memorial Bridge is 1772 m (5811 ft).

12. Let x mi be the length of the Indiana Toll Road, then $\frac{x}{3} - \frac{x}{4} = 13 \Rightarrow x = 156$ mi, and $\frac{156}{4} = 39$ mi. The length of the Massachusetts turnpike is $4(39 - 4.5) = 138$ mi. The Indiana Toll Road is 156 mi and the Massachusetts turnpike is 138 mi.

13. Only four numbers, 0, 1, 2, 3 squared result in a one-digit number. 0 and 1 does not work because then the third digit will be as the first or second digits. Hence, 2 and 3 are left, that give the digits 2^2, 3^2, $2 \cdot 3$ and the number 496.

The New York State Thruway is 496 mi.

14. The product of only two powers consequent numbers $2^2 \cdot 3^3 = 108$ can lead to a desirable result because $1^2 \cdot 2^2$ is 4, and $3^2 \cdot 4^2$ is already 6912, which is a four-digit number.

The Jane Addams Memorial Tollway is 108 mi.

15. Let x be the driving speed resulted in a speeding ticket, then $\frac{360}{x-8} - \frac{360}{x} = \frac{1}{2} \Rightarrow x^2 - 8x - 5760 = 0 \Rightarrow x = 80$. The second solution to the equation $x = -72$ does not satisfy the problem.

The allowed speed is 70 mph.

16. Let x mi be the length of I-90, then $\frac{x}{5} - \frac{x}{7} = 177.2 \Rightarrow \frac{2x}{35} = 177.2 \Rightarrow x = 3101$.

I-90 is 3101 mi.

17. Let 1 be the entire drive from Boston to Seattle. It takes Olga 5 days to drive, that is, she drives $\frac{1}{5}$ of the distance each day. Correspondingly, Vicki drives $\frac{1}{7}$ of the distance each day. When they drive together toward each other, they cover $\frac{1}{5} + \frac{1}{7}$ of I-90 per day. Taking x to be the number of days after Vicki left her home, we get $2 \cdot \frac{1}{5} + x\left(\frac{1}{5} + \frac{1}{7}\right) = 1 \Rightarrow \frac{14 + 12x}{35} = 1 \Rightarrow x = 1.75$.

The sisters met 1.75 days after Vicki left her house.

V

Constructions and Inventions in the United States

V.1 Architecture Wonders

Magnificent opera houses, fabulous concert halls, amazing lighthouses, animal-shaped buildings, and other US remarkable structures are among the tallest, oldest, or largest must-see constructions in the world.

1.–3. *The Metropolitan Opera House in New York City is one of the world's largest purpose-built opera houses with the most technologically advanced stages. Designed by American architect Wallace K. Harrison (1895–1981), it can seat 3800 visitors. Giacomo Puccini's opera La Fanciulla del West was the first opera performed there on April 11, 1966. However, the official opening of the Metropolitan Opera House was on September 16, 1966 with Samuel Barber's world premiere opera Antony and Cleopatra.*

1. The Metropolitan Opera's numbers of underground u and above-ground levels a, $u < a$, are co-primes and satisfy the equation: $\frac{1}{u} + \frac{1}{a} = \frac{14}{45}$. How many levels (stories) does the Metropolitan Opera House? How many of them are underground and above ground?

2. The lobby of the Metropolitan Opera House has two murals, *The Triumph of Music* and *The Sources of Music* by Marc Chagall (1887–1985). They were painted in France and shipped to New York. To prevent fading, the murals are covered during the hours of direct sunlight. The value of these paintings is estimated at $20 million.

   What are the dimensions of the murals? Use the minimal number of the hints to answer this question:

   1. The width of the mural is 6 ft less than its length.
   2. The area of the mural is 1080 sq. ft.
   3. The perimeter of the mural is 132 ft.

DOI: 10.1201/9781003229889-5

3. The centerpieces of the Metropolitan Opera House are the unique crystal chandeliers donated by the Vienna State Opera, Austria. They resemble constellations with sparkly moons and satellites. The largest one is 18 ft in the diameter. Some of the chandeliers in the auditorium are raised up during a performance.

   The auditorium has 10 more chandeliers than the lobby. The average number of chandeliers in the lobby and auditorium is 16. The average number of chandeliers in the auditorium that cannot be raised up and in the lobby is 10. How many crystal chandeliers are in the lobby and auditorium of the Metropolitan Opera House? How many of them can be raised up?

4. The Red Rock Amphitheatre near Morrison, Colorado, is a naturally occurring geological marvel with a huge disc-shaped rock behind the stage, a huge vertical rock, and several large outcrops. It has hosted many performances since its opening on May 31, 1906. The Red Rock Amphitheatre has been featured in several TV episodes and movies.

   The number of seats at the Red Rock Amphitheatre can be presented as the k-th term T_k, $k = 1, 2, \ldots, n + 1$ of $(\sqrt[8]{8} + 5)^n$ extended to the polynomial. The $(k - 1)$, k, and $(k + 1)$ terms are related to as $T_{k-1}/T_k/T_{k+1} = 2\sqrt[4]{8}/45\sqrt[8]{8}/600$, $k = 2, \ldots, n$. How many seats does the theater have?

5. The Chrysler and Empire State Buildings in New York City used to be the tallest buildings in the world. Decorative elements of the Chrysler Building resemble parts of a car, while the Empire State Building reminds ancient pyramids.

   The Empire State Building is 204 ft taller than the Chrysler Building. Their heights can be presented as the products of ab^2 and ad, respectively, where a, b, and d are positive integers. The sum $a + d$ is 21 times b. The difference between b^2 and $a + d$ is 25 times a^2. The number of levels in Empire State Building is $a(2b + 1)$ and in the Chrysler Building is $3b + a$. How tall are these buildings? How many levels do they have?

6.–8. *The Absecon and Barnegat Lighthouses in New Jersey, Cape Hatteras and Cape Lookout Lighthouses in North Carolina, Cape Henry Lighthouse in Virginia, and St. Augustine Lighthouse in Florida are among the tallest lighthouses in the United States that were built to be lighthouses.*

6. The ranking in the list of the tallest lighthouses that Absecon Lighthouse and Cape Lookout Lighthouse have coincide with the smallest and largest distances between any two points of the curve presented by $9x^2 + 25y^2 - 72x - 50y - 56 = 0$. The ranking of Barnegat

Lighthouse and Cape Hatteras Lighthouse are the x and $y$coordinates of its center. The ranking of St. Augustine Lighthouse and Cape Henry Lighthouse are x-coordinates of one of its foci and vertices. What ranking in the list of the tallest lighthouses do the Absecon, Barnegat, Cape Hatteras, Cape Henry, Cape Lookout, and St. Augustine lighthouses have? Describe the curve defined by this equation.

7. The St. Augustine Lighthouse is the eighth tallest lighthouse in the USA. It is located on the north end of Anastasia Island in St. Augustine, Florida. Visitors can climb its spiral staircases to the top of the St. Augustine Lighthouse.

   Philip and Andrew met at the top of the St. Augustine Lighthouse. Enjoying the gorgeous panoramic views and talking about their climbing experiences, Philip and Andrew found that they both made three stops with the third stop at the top. The numbers of steps they covered between the first and second stops were the same. However, the numbers of steps between three stops covered by Philip are terms of an arithmetic sequence, and by Andrew are terms of a geometric sequence. The sum of the first five terms of the geometric sequence is 365. How many steps does the St. Augustine Lighthouse have?

8. The gem of Cape Hatteras National Seashore, the Cape Hatteras Lighthouse, was built in 1870 by the US Army Corps of Engineers and moved to its present location in 2000. The Cape Hatteras Lighthouse is the tallest lighthouse in the nation built to be a lighthouse.

   The three-digit number of steps visitors should climb to the top of the Cape Hatteras Lighthouse can be designed by means three consecutive integers. The product of the first two integers is the first digit of the number of steps. The product of the second and third integers is its second digit. The second integer raised to the third integer is third digit (units) of the number of steps.

   The same three integers can determine the height in feet. The value of the height of the Cape Hatteras Lighthouse from the foundation to the top of the roof spire is the product of three factors, the second integer, the number formed by the first and third integers (in this order), and the second integer raised to the third integer.

   How many steps does the Cape Hatteras Lighthouse have? How tall if the lighthouse?

9.–11. *The Newport Tower in Touro Park, Rhode Island, is surrounded by a historic residential neighborhood. It is also called the Round Tower, Touro Tower, Newport Stone Tower, and Old Stone Mill, though the tower is not exactly circular by several* *inches. The tower stands on a hill above a waterfront tourist district. The 18th-century paintings show that the tower used to have an amazing view of the harbor.*

9. The cylindrical walls of the Round Tower are 3-ft thick. What statements below are correct?

   A. The volume of the tower (exterior structure) is 3 cubic ft greater than the volume of the inner chamber.

   B. The volume of the tower is 9 cubic ft greater than the volume of the inner chamber.

   C. The volume of the tower is 27 cubic ft greater than the volume of the inner chamber.

   D. The area of the exterior circular base of the tower is 3 sq. ft greater than its inner circular base.

   E. The area of the exterior circular base of the tower is 9 sq. ft greater than its inner circular base.

   F. The area of the exterior circular base of the tower is 27 sq. ft greater than its inner circular base.

   G. Not enough information to answer the question.

10. Let us imagine that a triangular fence is constructed inscribing the circular base of the Round Tower. The areas of three triangles formed by connecting the center of the circular base with vertices of the triangle are 216, 288, and 360 sq. ft. Find the diameter of the Newport Tower and the sides of a triangular fence.t

11. The Round Tower is 28-ft tall with the inner chamber diameter of 18 ft and 3-ft thick walls. What is the volume of the stone structure (the tower wall)?

12. *The Big Bull, Giant Koala Building, and the Crocodile Hotel* in Australia, *the Crab Building* in the Philippines, *the Conch Shell House* in Mexico, *the Elephant Tower* in Thailand, and *the Whale Building* in Japan are just a few examples of gorgeous animal-shaped buildings. The USA is also home to such unique structures. The restaurant *Longhorn Grill* in Amado, Arizona, and the KFC restaurant *The Big Chicken* in Marietta, Georgia, not only serve delicious food, but are also world-famous animal-shaped buildings.

    How tall are the Longhorn Grill and Big Chicken if their heights are a and b in the quadratic equation

    $$2x^2 - ax + 2b = 0$$

    that has the roots of 7 and 8?

13. Buildings of the Fresh Water Fishing Hall of Fame in Hayward, Wisconsin, founded in 1960, are dedicated to promoting freshwater fishing. One of the buildings is a fiberglass sculpture of a jumping muskie fish. The lower jaw of the fish is an observation deck. The Hall of Fame keeps the records for the largest freshwater fish in the USA and the world.

   The length of the fiberglass sculpture is an integer value a such that the quadratic equation

   $$\frac{a-23}{30}x^2 + 53 = \frac{a-31}{2}x - a$$

   has just one solution. Moreover, this solution represents the number of buildings in the Fresh Water Fishing Hall of Fame. What is the length of the fiberglass sculpture? How many buildings are in the Fresh Water Fishing Hall of Fame?

14. Paul Bunyan is a giant lumberjack in American folklore that performed superhuman tasks. The advertising manager for the Red River Lumber Company William B. Laughead brought Paul with his Babe the Blue Ox to life in a 1916 promotional pamphlet. Paul Bunyan then became a main character in the 1958 Walt Disney musical. Statues in his honor can be found in Akeley, Bemidji, and Brainerd (Minnesota), Manistique and Ossineke (Michigan), Muncie (Indiana), Lakewood (Wisconsin), and other places across the USA.

   Two famous statues of Paul Bunyan of the same height were erected in 1959. The first one, near Portland, Oregon, is listed in the National Register of Historic Places. The other one, built in Bangor, Maine, is the world's largest statue of Paul Bunyan. Klamath, California, is proud of its Statue of Paul Bunyan as well, though it is not as tall as the one in Maine. The height of the statues in California and Oregon are the second and fifth terms of an arithmetic sequence in which the first term is related to the sixth as 11 to 5. The ninth term squared is equal to the second term. How tall are the Paul Bunyan Statues in California, Maine, and Oregon?

15. Located at an elevation above 7000 ft in eastern Arizona and completed in 1991, the Round Valley Ensphere was the first domed high school football stadium in the country. It costs $11 million. The Round Valley Ensphere is listed as number 1 among the *Ten High School Football Stadiums to See Before You Die*.

The Round Valley Ensphere has the floor area of 189,000 sq. ft and a diameter of 440 ft. Does the circular ground area of the Round Valley Ensphere include anything but the area below the dome? Assuming a hemisphere dome, what would be the volume of the structure?

Answers

1. 14, 5, 9
2. 30 ft x 36 ft, 2
3. 11, 21, 12
4. 9000
5. 1250 ft, 1046 ft, 102, 77
6. 6, 4, 1, 9, 8, an ellipse
7. 219
8. 268 steps, 208 ft
9. G
10. 24 ft, 36, 48, 60 ft
11. 5542 cubic ft
12. 30 ft, 56 ft
13. 143 ft, 7
14. 31 ft, 49 ft
15. yes, 22,301,119 cubic ft

Solutions

1. Since u and a are co-primes, from $\frac{1}{u} + \frac{1}{a} = \frac{u+a}{ua} = \frac{14}{45}$ follows $ua = 45$ and $u + a = 14$, which give the number of stories of $14 = u + a$.

   1st way. Applying the Viète Theorem (see Appendix I-QE), we obtain $u = 5$ and $a = 9$ or $u = 9$ and $a = 5$.

   2nd way. Expressing $a = 14 - u$ from the second equation and substituting it to the first equation, we obtain the quadratic equation $u(14 - u) = 45$ or $u^2 - 14u + 45 = 0$. Its solutions (see Appendix I-QE) are $u = 5$ or $u = 9$. Then $a = 9$ and $a = 5$ correspondingly. Because $u < a$, $u = 5$ and $a = 9$ are taken.

   There are 5 underground and 9 above ground levels in the Metropolitan Opera House.

2. Let w and l be the width and length of the mural, then $l - m = 6$ (Hint 1), $2l + 2m = 132$ (Hint 3), $lm = 1080$ (Hint 2). Combining any two hints, we get $l = 36$, $w = 30$.

Two murals in the Metropolitan Opera House are 30 ft by 36 ft.

3. Let l and a be the numbers chandeliers in the lobby and auditorium, then $a - l = 10$ and $(l + a)/2 = 16$. Therefore, $a = 21, l = 11$. Then, the number of chandeliers that can be raised up is found from $(x + 11)/2 = 10$ as $x = 9$ and then $21 - 9 = 12$.

The Metropolitan Opera House has 11 chandeliers in the lobby and 21 in auditorium, 12 of them can be raised up.

4. Applying the binomial expansion (see Appendix I-AF) to $(\sqrt[8]{8} + 5)^n$, we obtain the polynomial

$$C_n^0(\sqrt[8]{8})^n \cdot 5^0 + C_n^1(\sqrt[8]{8})^{n-1} \cdot 5^1 + \ldots + + \ldots + C_n^n(\sqrt[8]{8})^0 \cdot 5^n.$$ The polynomial has $n + 1$ terms, $T_k = C_n^{k-1}(\sqrt[8]{8})^{n-(k-1)} \cdot 5^{k-1}$, $k = 1, n + 1$. Then

$$T_{k-1}/T_k/T_{k+1} = (C_n^{k-2}(\sqrt[8]{8})^{n-(k-2)} \cdot 5^{k-2})/(C_n^{k-1}(\sqrt[8]{8})^{n-(k-1)} \cdot 5^{k-1})/(C_n^k$$

$$(\sqrt[8]{8})^{n-k} \cdot 5^k) = \left(\frac{n!}{(k-2)!\,(n-k+2)!}(\sqrt[8]{8})^{n-(k-2)} \cdot 5^{k-2}\right)/\left(\frac{n!}{(k-1)!\,(n-k+1)!}(\sqrt[8]{8})^{n-(k-1)} \cdot 5^{k-1}\right)/$$

$$\left(\frac{n!}{k!\,(n-k)!}(\sqrt[8]{8})^{n-k} \cdot 5^k\right),$$ that is the same as $2\sqrt[4]{8}/45\sqrt[8]{8}/600$. Therefore, we can present these ratios as the system:

$$\begin{cases} \frac{(\sqrt[8]{8})^2(k-1)(n-k+1)}{(n-k+2)(n-k+1)\cdot\sqrt[8]{8}\cdot 5} = \frac{2\sqrt[4]{8}}{45\sqrt[8]{8}} \\ \frac{\sqrt[8]{8}\cdot 5(k-1)k}{(k-1)(n-k+1)\cdot 25} = \frac{45\sqrt[8]{8}}{600} \end{cases}$$ that can be simplified to $\begin{cases} \frac{k-1}{n-k+2} = \frac{2}{9} \\ \frac{k}{n-k+1} = \frac{3}{8} \end{cases}$.

The solutions to the system $k = 3$ and $n = 10$ determine the term $T_3 = C_{10}^{3-1}(\sqrt[8]{8})^{10-(3-1)} \cdot 5^{3-1} = \frac{10!}{2!\cdot 8!} \cdot 8 \cdot 25 = 9000$.

There are 9000 seats in the Red Rock Amphitheatre.

5. As follows from the problem, $ab^2 - ad = 204$, $a + d = 21b$, and $b^2 - (a + d) = 25a^2$. Multiplying both sides of the third equation by a and using the first equation, we get $25a^3 + a^2 = 204$ or $a^2(25a + 1) = 2^2 \cdot 51$. It is easy to see that the only integer solution is $a = 2$. Then substituting the second equation $a + d = 21b$ to the third $b^2 - (a + d) = 25a^2$, we get a quadratic equation with respect to b: $b^2 - 21b - 100 = 0$ that has two solutions, $b = 25$ and $b = -4$. The second solution does not satisfy the applied meaning of the problem. Then, $d = 523$.

The Empire State Building has 102 stories, and the Chrysler Building has 77.

6. From $9x^2 + 25y^2 - 72x - 50y - 56 = 0$ follows $9(x^2 - 2 \cdot 4x + 4^2) + 25(y^2 - 2 \cdot 1y + 1^2) = 225$, which is simplified to $9(x - 4)^2 + 25(y - 1)^2 - 225 = 0 \Rightarrow \frac{(x-4)^2}{5^2} + \frac{(y-1)^2}{3^2} = 1$. The last equation is the equation of an ellipse (see Appendix I-QC) with the center $(4, 1)$, foci $(0, 1)$ and $(8, 0)$,

vertices $(-1, 1)$ and $(9, 1)$, co-vertices $(4, -2)$ and $(4, 4)$.

The Absecon Lighthouse and Cape Lookout Lighthouse are the sixth (3×2) and tenth (5×2) tallest lighthouses, the Barnegat Lighthouse and Cape Hatteras Lighthouse are the fourth tallest and the tallest lighthouses, the St. Augustine Lighthouse and Cape Henry Lighthouse are the eighth and ninth tallest lighthouses.

7. Let a be the middle number of both sequences and d and q be the difference and ratio of the arithmetic and geometric sequences (see Appendix I-SS), then $(a - d) + a + (a + d) = \frac{a}{q} + a + aq \Rightarrow q^2 - 2q + 1 = 0 \Rightarrow q = 1$, which implies that all terms of the geometric sequence are the same. Then from $a = 365/5$ follows $a = 73$ and $73 \cdot 3 = 219$.

219 steps lead to the top of the St. Augustine Lighthouse.

8. Let us test 1, 2, 3, then the number of steps is 268 $(1 \cdot 2, 2 \cdot 3, 2^3)$ and $2 \cdot 13 \cdot 2^3 = 208$ ft. The next sequence of consecutive integers 2, 3, 4 does not work because 3^4 is 12, which is not a digit.

268 steps lead to the top of the 208-ft-tall Cape Hatteras Lighthouse.

9. Let R and r be the circumradius and inradius and H be the height of the tower. Then $R - r = 3$ ft. The volume of the stone structure is the difference of volumes of the tower and its inside: $V = \pi R^2 H - \pi r^2 H = \pi H(R^2 - r^2) = \pi H(R - r)(R + r) = 84\pi(R + r)$ cubic ft (see Appendix I-VS). The difference of the areas is $A = \pi R^2 - \pi r^2 = \pi(R^2 - r^2) = 3\pi(R + r)$. The value of $R + r$ is not provided.

It is not enough information to answer the question based on the information provided.

10. Let us check whether the triangle is a right triangle. Yes, it is. Indeed, $216^2 + 288^2 = 360^2$. Using the formulas for the area of a triangle, $A_a = \frac{ar}{2}$, $A_b = \frac{br}{2}$, $A_c = \frac{cr}{2}$, and $A_a^2 + A_b^2 = A_c^2$, we get $\left(\frac{r}{2}\right)^2 (a^2 + b^2) = \left(\frac{r}{2}\right)^2 c^2$, where a, b, and c are sides of the triangle, r is the inradius which is perpendicular to each side, and A_a, A_b, and A_c are the areas of the three triangles with the corresponding side as the base. Let c be the hypotenuse. The area inside the triangular fence is $A = 216 + 288 + 360 = 864$. On the other side, the area of the right triangle is $A = \frac{ab}{2} = \frac{4A_a A_b}{2rr}$, that is, $864 = \frac{2 \cdot 216 \cdot 288}{r^2}$, from which $r = 12$. Using the formulas for the area, we get $a = 36$, $b = 48$, $c = 60$.

The diameter of the Round Tower is 24 ft.

11. The radii of the exterior and interior circles and the height of the tower are $R = 12$ ft, $r = 9$ ft, and $H = 28$ ft. Then the volume of the stone structure is the difference between volumes of two cylinders: $V = \pi R^2 - \pi r^2 H = 1764\pi \approx 5541.77$ cubic ft.

The volume of the stone structure of the Round Tower is 5542 cubic ft.

12. In the quadratic equation $2x^2 - ax + 2b = 0$ or $x^2 - ax/2 + b = 0$, the product of its solutions is $7 \cdot 8 = 56 = b$. The sum of its solutions is $7 + 8 = 15 = a/2$. Hence, $b = 56$ and $a = 30$.

The Longhorn Grill is 30 ft tall and Big Chicken is 56 ft tall.

13. The quadratic equation $\frac{a-23}{30}x^2 + 53 = \frac{a-31}{2}x - a$ can be rewritten as $\frac{a-23}{30}x^2 - \frac{a-31}{2}x + (53 + a) = 0$. It has a unique solution when its discriminant (see Appendix I-QE) $\left(\frac{a-31}{2}\right)^2 - 4 \cdot \frac{a-23}{30}(53 + a) = 0$ or $7a^2 - 1170a + 24167 = 0$, that leads to $a = 143$. The other solution, $a = 169/7$, is not integer and, thus, does not work.

Then, the quadratic equation $\frac{a-23}{30}x^2 + 53 = \frac{a-31}{2}x - a$ becomes $4x^2 - 56x + 196 = 0$, or $(2x - 14)^2 = 0$, with the solution $x = 7$.

There are 7 buildings in the Fresh Water Fishing Hall of Fame. The fiberglass sculpture centerpiece is 169/7 ft long.

14. Let a and d be the first term and difference of the arithmetic sequence (see Appendix I-SS). From the combination of two conditions $\frac{a_1}{a_6} = \frac{a}{a + 5d} = \frac{11}{5}$ follows $a = -\frac{55}{6}d$, that with $a_9^2 = a_2$ or $(a + 8d)^2 = a + d$ leads to the quadratic equation $\left(-\frac{55}{6}d + 8d\right)^2 = -\frac{55}{6}d + d$. The equation is simplified to $\left(-\frac{7}{6}d\right)^2 = -\frac{49}{6}d$, that has two solutions, $d = 0$ and $d = -6$. The solution to the quadratic equation $d = 0$ does not work because then all statues would have the same height. Then, $d = -6$ and $a = 55$. Hence, $a_2 = 49$ ft and $a_5 = 31$ ft.

The Paul Bunyan Statues in California and Oregon are 49 ft and 51 ft tall correspondingly.

15. The area of the circular floor with the radius of 220 ft is $220^2\pi = 152{,}053.08$ sq. ft, which is less than the actual area of 189,000 sq. ft. Therefore, it includes some additional rooms. The volume of the structure is $\frac{1}{2} \cdot \frac{4}{3} \cdot 220^3\pi = 22{,}301{,}119$ cubic ft (see Appendix I-VS).

The volume of the Round Valley Ensphere is 22,301,119 cubic ft.

V.2 Incredible Constructions

New constructions have been continuously designed and built. Some of them become the largest, the tallest, the longest, the heaviest, or the most expensive. They keep their title for quite some time.

1. The *most expensive* project, the *Interstate Highway System* in the USA, was proposed by the 34th US President Dwight David Eisenhower. Being a general, he understood the importance of well-developed connections for the nation and the military.

   Construction of new highways and improvements of old roads for Interstate Highway System cost billions of dollars. This amount and the year the project began have 17 as a common factor, while 4 is the quotient and 119 is the remainder after the year is divided by the cost value (in billions of dollars). The year when the project began is also divisible by 5 and 23. How expensive is the Interstate Highway System project? When did the project begin? Are all statements of the problem needed to answer these questions?

2. *Big Dig* or the *Central Artery/Tunnel* Project in Boston is the *most expensive* highway project not only in the USA but also in the world. Its development started in 1982 and construction began in 1991 with the estimated completion in 1998 that was delayed until 2007, when it was finally completed at a larger cost than initially planned.

   The estimated and actual costs are the third and eighth terms of an arithmetic sequence with 14 as the sum of its first five terms and 88 as the sum of its first ten terms. What are the estimated and actual costs in billion dollars?

3.–5. *Cloud Gate or "The Bean" at Millennium Park, Chicago, Illinois, was designed by Indian-born British artist Anish Kapoor and constructed between 2004 and 2006. All funding for this project came from donations.*

3. It was initially estimated that the Bean would cost $6 million. This amount was increased by 91.67% in 2004, then was increased by 100% more by 2006. What are the estimated cost of the *Bean* in 2004 and its final cost? What is the increase of the final cost compared to the initially estimated cost?

4. The Bean is made from $\dfrac{1}{\left(\frac{1}{13}\right)^2 + \left(\frac{1}{13}\right)^4 + \left(\frac{1}{13}\right)^6 + \left(\frac{1}{13}\right)^8 + \ldots}$ stainless-steel plates that were polished to avoid any visible seams. How many stainless-steel plates are used?

5. If the Bean were 2 ft taller, then its width, height, and length would be the third, fourth, and sixth terms of an arithmetic sequence. If the Bean were 13 ft taller, then it would be the fifth terms of the same arithmetic sequence. The first term and the difference of the arithmetic sequence are the same. What are the width, length, and height of the Bean?

6. The Washington Monument has been the *tallest* non-communication structure in the District of Columbia since its completion in 1884,

and was the *tallest* monument in the USA until the San Jacinto Monument was erected in Texas in 1939.

The Washington Monument was the *tallest* structure in the world for the number of years equaled the maximum value y of the function $y = 3 \sin 2x + 4 \cos 2x$. Its height in feet also presented by y written back-to-back three times (e.g., 6 written back-to-back three times is 666). How long was the Washington Monument the tallest structure in the world? How tall is it?

7. The Chrysler Building and Empire State Building are notable Manhattan structures. The Chrysler Building was built for auto-motive magnate Walter P. Chrysler in deco-style with automobile parts in its design. The Empire State Building reminds an ancient pyramid with a mast at the top. Both famous buildings used to be the *tallest buildings* in the world.

The Chrysler Building was built in 1930 and was the world's tallest building for x years until the Empire State Building was built and became the tallest for y years, where the pair (x, y) represent the maximum value of the function $y = 35 - x^2 + 2x$. When was the Empire State Building built? How long was it the tallest?

8. The Palm Springs Aerial Tramway in California is the *largest* ro-tating aerial tram in the world. Since its opening in September 1963, it has lifted thousands of visitors from the Coachella Valley at 2643 ft up to the Mountain Station at 8516 ft. Rotating constantly, aerial cars pass through different life-zones providing visitors with a 360° panoramic view.

The number of rotations the tram makes and the time in minutes it takes to go up are the maximum (x, y) of a continuous odd function $y = f(x)$. The function $y = f(x)$ is defined on a set of all real numbers, has the minimum at $(-2, -12.5)$, and a horizontal asymptote $y = -5$. How many bio-zones does a tram pass if this number is the value that $y = f(x)$ approaches as x approaches infinity? How many rotations does it makes? What is the time of one ride? How many other rotating aerial trams are in the USA if this number is the maximum number of vertical asymptotes the function $y = f(x)$ may have? Is continuity of the function $y = f(x)$ important to solve the problem? Is the fact that the function $y = f(x)$ is odd important to solve the problem?

9. The longer of the two parallel bridges of the Lake Pontchartrain Causeway that cross Lake Pontchartrain in Louisiana was listed by *Guinness World Records* in 1969 as the *longest* bridge over water in the world.

One car starts driving the bridge at 40 mph. The second car that drives 8 mph faster enters the bridge 5 minutes after the first car but exits the bridge one minutes before the first car. How long is the bridge?

10.–14. *The heaviest and the largest balls of twine are in the USA. Francis A. Johnson from Minnesota, Frank Stoeber from Kansas, James Frank Kotera from Wisconsin, and J. C. Payne from Texas made the most remarkable twine balls at different years.*

10. The ball of twine started in 1953 weighs 486 lb more than the ball of twine started in 1950. The total weight of balls of twine made in 1987 and 1950 is 29,400 lb. The ball of twine made in 1950 weighs 2900 lb more than 10 times the difference between the balls started in 1979 and 1953 (lighter). The ball of twine started in 1979 weighs 4664 lb less than double the twine ball started in 1987. Find the weight of each balls of twine, the mean and median of these balls.

11. Francis A. Johnson from Minnesota wrapped twine to form his ball for 4 hours each day. If he spent only 3 hours each day for those years with the same productivity, what would be the diameter of his ball of twine?

   Use the following information is necessary to solve the problem:

   A. Francis A. Johnson wrapped twine to form his ball for 29 years.

   B. The Johnson's twine ball is referred to as *The Pride of Darwin,* Minnesota.

   C. The Johnson's twine ball is considered the *World's Largest Ball of Twine Built By One Man.*

   D. The final diameter of the ball is 12 ft.

12. If the twine that Francis A. Johnson from Minnesota used to wrap his ball with the diameter of 12 ft were to put into a cylinder with the base diameter of 12 ft, what would be the height of the smallest cylinder to hold the twine? Select the answer

A	6 ft
B	8 ft
C	10 ft
D	12 ft

13. If the twine that Francis A. Johnson from Minnesota used to wrap his ball with a diameter of 12 ft were to be put into a box with a square base with the side of 12 ft, what would be the height of the smallest box to hold the twine?

14. If the twine ball with the diameter of 12 ft made by that Francis A. Johnson from Minnesota were to be put into a cone with the height equaled the diameter of the base. What would be the radius of the base and the volume of the cone?

Answers

1. $459 billion, 1955, no
2. $2.8 billion, $14.8 billion
3. $11.5 million, $23 million, 383.33%
4. 168
5. 33 × 66 × 42
6. 5, 555 ft
7. 1931, 36 years
8. 5, 2, 12.5 min, 0, no, yes
9. 24 mi
10. 1950: 17,400 lb, 1953: 17,886 lb, 1979: 19,336 lb, 1987: 12,000 lb, 16,655.5 lb, 17,643 lb
11. $6\sqrt[3]{6}$, D
12. B
13. 6.28 ft
14. $R = 3(1 + \sqrt{5})$ ft; $V = 144\pi(2 + \sqrt{5})$ cubic ft

Solutions

1. The year when the project began is divisible at least by 5, 17, and 23, which are primes and co-primes as well (see Appendix I-PF). Multiplying these numbers, we get 1955. So, it is unnecessary to look for other factors for the year. The amount (in billions of dollars) is obtained from $1955 = 4x + 119$ as 459. The fact that this amount is divisible by 17 is not used.

The Interstate Highway System project began in 1955 and cost $459 billion.

2. The sum of the first n terms of an arithmetic sequence is $S_n = \frac{2a + d(n-1)}{2}n$ (see Appendix I-SS), then $S_5 = \frac{2a + d(5-1)}{2} \cdot 5 = 14$ and $S_{10} = \frac{2a + d(10-1)}{2} \cdot 10 = 88$, which leads to the system
$$\begin{cases} 5(2a + 4d) = 28 \\ 5(2a + 9d) = 88 \end{cases} \Rightarrow \begin{matrix} a = -2 \\ d = 2.4 \end{matrix} \Rightarrow \begin{matrix} a_3 = -2 + 2 \cdot 2.4 = 2.8 \\ a_8 = -2 + 7 \cdot 2.4 = 14.8 \end{matrix}.$$

The estimated and actual costs of Big Dig are $2.8 billion and $14.8 billion.

3. 91.67% increase of $6 million $= \frac{6 \cdot 191.67}{100} = 11.5$ million dollars. 100% increase of this amount is $23 million or $\frac{23 \cdot 100}{6} = 383.33\%$.

There is 383.33% increase of estimated costs for *The Bean* construction.

4. $\left(\frac{1}{13}\right)^2 + \left(\frac{1}{13}\right)^4 + \left(\frac{1}{13}\right)^6 + \left(\frac{1}{13}\right)^8 + \dots$ is an infinite geometric series with the first term of $\left(\frac{1}{13}\right)^2$ and the ratio of $\left(\frac{1}{13}\right)^2$, which is less than 1 (see Appendix I-SS). Then its sum is $\frac{\left(\frac{1}{13}\right)^2}{1-\left(\frac{1}{13}\right)^2} = \frac{1}{168}$.

The Bean consists of 168 stainless-steel plates.

5. Let a_i denote the i-th term of an arithmetic sequence and h is the height. Then $a_5 - a_4 = (h + 13) - (h + 2) = 11$, which is the difference of the arithmetic sequence and its first term. Then $33 = 11 + 2 \cdot 11$, $44 - 2$, 66.

The Bean dimensions are 33 x 66 x 42.

6. $y = 3\sin 2x + 4\cos 2x \Rightarrow y = 5 \cdot \left(\frac{3}{5}\sin 2x + \frac{4}{5}\cos 2x\right)$. Since $\left(\frac{3}{5}\right)^2 + \left(\frac{4}{5}\right)^2 = 1$, then there is an angle φ such that $\sin\varphi = \frac{3}{5}$ and $\cos\varphi = \frac{4}{5}$. The equation for y can be rewritten (see Appendix I-TF) as $y = 5 \cdot (\sin\varphi\sin 2x + \cos\varphi\cos 2x)5\cos(\varphi - 2x)$. The inequality $-1 \le \cos\alpha \le 1$, where α is any angles, implies $-5 \le 5\cos(\varphi - 2x) \le 5$ that brings 5 as the maximum value for the function $y = 3\sin 2x + 4\cos 2x$.

With the height of 555 ft, the Washington Monument was the tallest structure in the world for 5 years.

7. The function $y = 35 - x^2 + 2x$ is simplified to $y = -(x - 1)^2 + 36$. It is the equation of a parabola that opens downward (see Appendix I-QC). Its maximum is at (1, 36).

The Chrysler Building was the world's tallest building for one year. The Empire State Building was built and became the tallest in 1931 and kept this title for 36 years.

8. The function $y = f(x)$ is odd, then, regardless of continuity or discontinuity, it is symmetric about the origin. The point (2, 12.5) is the point of maximum (because (−2, −12.5) is the minimum), that gives 2 rotations during 12.5-minute ride. The line $y = 5$ is another horizontal asymptote. Hence, $y = f(x)$ approaches 5 as x approaches

infinity. There are no vertical asymptotes because the function is defined at a set of all real numbers. Thus, there is no other rotating aerial trams in the USA.

The Palm Springs Aerial Tramway passes 5 bio-zones during its 12.5-minute ride.

9. Let x be the length of the bridge, then $\frac{x}{40}$h is the time needed for the first car to pass the bridge, $\frac{x}{48}$h is the time needed for the second car to pass the bridge, and $\frac{x}{40} = \frac{5}{60} + \frac{x}{48} + \frac{1}{60} \Rightarrow \frac{x}{40} - \frac{x}{48} = \frac{1}{10} \Rightarrow x = 24$.

The Lake Pontchartrain Causeway is a 24-mi-long bridge.

10. Let the subscript of the weight x of the ball be the last digit of the year, then

$$
\begin{cases}
x_3 - x_0 = 486 \\
x_7 + x_0 = 29400 \\
x_0 = 10(x_9 - x_3) + 2900 \\
x_9 = 2x_7 - 4664
\end{cases}
\text{or}
\begin{cases}
x_3 = x_0 + 486 \\
x_7 = -x_0 + 29400 \\
x_0 = 10((-2x_0 + 54136) - (x_0 + 486)) + 2900 \\
x_9 = -2x_0 + 54136
\end{cases}
$$

Hence, $31x_0 = 539400 \Rightarrow x_0 = 17400$. Thus, $x_3 = 17{,}886$ lb, $x_9 = 19{,}336$ lb, $x_7 = 12{,}000$ lb.

The mean: $(19{,}336 + 17{,}400 + 17{,}886 + 12{,}000)/4 = 16{,}655.5$. The median is the average of two middle numbers, which is $(17{,}400 + 17{,}886)/2 = 17{,}643$.

The weight of the balls with the year they were made are 1950: 17,400 lb, 1953: 17,886 lb, 1979: 19,336 lb, 1987: 12,000 lb.

11. Because the same amount of volume is added to the ball each day, then from $V_{ball} = \frac{4}{3}\pi r^3$, we conclude that 4 hours is related to 6^3 as 3 hours is related to r^3, that is $\frac{6^3}{4} = \frac{r^3}{3}$. Then $r^3 = \frac{6^3 \cdot 3}{4} = 162$ and $r = 3\sqrt[3]{6}$. Hint (D) is used.

The diameter of Francis A. Johnson's ball of twine would be $6\sqrt[3]{6}$.

12. Using $\frac{4}{3}\pi r^3 = V_{ball} = V_{cyl} = \pi r^2 h$, we get $\frac{4}{3}r = h$ from which follows $h = \frac{4}{3}r$ or $h = 8$ at $r = 6$.

The height of a cyllender to fit Francis A. Johnson's ball of twine should be 8 ft.

13. $r_{ball} = 6$; $V_{ball} = V_{box} \Rightarrow V_{ball} = \frac{4}{3}\pi r^3$, $V_{box} = (2r)^2 h \Rightarrow h = \frac{\pi}{3}r = 6.28$.

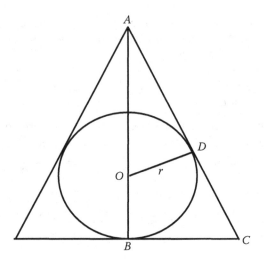

FIGURE V.1

The height of a box to fit Francis A. Johnson's ball of twine should
be 6.28 ft.

14. Let r be a radius of the twine ball, R and H be a radius of the base
and a height of the cone, shown in Figure V.1.

The triangles $\triangle ABC$ and $\triangle ADO$ are similar. Then $OD = r = 6$,
$BC = R$; $AB = H = 2R$.

Hence, $\frac{H-r}{r} = \frac{\sqrt{H^2 + R^2}}{R} \Rightarrow 2R = r + r\sqrt{5} \Rightarrow R = r\frac{1+\sqrt{5}}{2} = 3(1 + \sqrt{5})$.

$$V_{conw} = \frac{1}{3}\pi R^2 H = \frac{2}{3}\pi R^3 = \frac{2}{3}\pi (3(1 + \sqrt{5}))^3 = 144\pi (2 + \sqrt{5}).$$

The volume of a cone to fit Johnson's ball of twine would be
$144\pi (2 + \sqrt{5})$ cubic ft.

V.3 Predictions and Reality

Each generation believes that it lives at the "the golden age. Everything that can be
invented has been invented" and "the invention of which long-ago reached its
limit", as the Commissioner of US patent office Charles H. Duell (the first quote)
stated in 1899 and the governor of Britannia (the second quote) Sextus Julius

Frontinus said almost two millenniums ago, in 84. The time has significantly adjusted their predictions. As English actor Peter Ustinov (1921–2004) commented, "if the world should blow itself up, the last audible voice would be that of an expert saying it can't be done".

1. Novel technological ideas have not always been appreciated. In his letter of January 31 to the seventh US President Andrew Jackson, New York State Governor Martin Van Buren, who later became the eighth US President, wrote: *The canal system of this country is being threatened by the spread of a new form of transportation known as 'railroads'. ... 'railroad' carriages are pulled at the enormous speed of fifteen miles per hour by 'engines' which, in addition to endangering life and limb of passengers, roar and snort their way through the countryside, setting fire to crops, scaring the livestock and frightening women and children... people should not travel at such breakneck speed.*

   What year was the Van Buren's letter dated if the sum of all digits of the year is 2 more than the product of its last two digits? In addition, the sum of the first two digits of the year is its last digit (units). The sum of the first and last digits is the same as the sum of the second and third digits. The second digit is not less than four times the third (tens) digit. None of the digits is repeated.

2. The First Transcontinental Railway, known as the *Pacific Railroad* or *Overland Route* was opened on May 10 of the year presented as the product of the smallest positive integers x and y such that two middle terms of $(x + y)^n$ are related to as 89 to 21 and n is odd. In 100 years, on July 20, the United States' Apollo-11 was the first manned mission to land on the Moon. What years were the First Transcontinental Railway open and the first manned mission landed on the Moon?

3. The First Transcontinental Railway connects the Pacific coast at San Francisco Bay with the Eastern US rail network at Council Bluffs, Iowa, making transportation from coast to coast much quicker and cheaper.

   The First Transcontinental Railway was built by three private companies. The parts of the rail line built by the Western Pacific Railroad Company and Central Pacific Railroad Company of California are related to as 22 to 115. The lengths built by the Central Pacific Railroad Company and the Union Pacific Railroad Company are related to as 138 to 217. The Central Pacific Railroad Company built

30 mi more than five times the length built by the Western Pacific Railroad Company.

What is the length of the First Transcontinental Railway?

A. 822 mi

B. 1217 mi

C. 1261 mi

D. 1907 mi

E. 2305 mi

4. A president of the Michigan Savings Bank advised Henry Ford's lawyer Rackham not to invest into automobile industry: *"The horse is here to stay, but the automobile is only a novelty—a fad".* Mr. Rackham ignored the advice and made investment in Ford stock, selling it later for the profit of $12.495 million. How much did Mr. Rackham invest in Ford stock if the selling amount was 2500 times the initial amount?

5. In his speech to the Aero Club of France on November 5, 1908, American aviation pioneer Wilbur Wright (1867–1912) stated: *"I confess that in 1901 I said to my brother Orville that man would not fly for fifty years. Just x years later we ourselves made flights. This demonstration of my impotence as a prophet gave me such a shock that ever since I have distrusted myself and avoided all predictions".*

In how many years after his prediction in 1901 did Wright brothers fly if the number x satisfies the equation

$$x^2 - 4x + 9 = 5\cos(x - 2)?$$

6. On January 13, the *New York Times* stated that *A rocket will never be able to leave the Earth's atmosphere* but withdrew their statement 33 years later, on July 17, when Apollo started its journey to the Moon. If the year of Apollo journey to the Moon is subtracted from the predicted year written upside down, then the difference would be 4992. What year did Apollo leave for the Moon? What year was the prediction made?

7. The Apollo missions that made the first manned flight to the Moon on December 21–27, 1968 and the first man landing on the Moon on July 16–24, 1969, have numbers that are the sum and product of all solutions to $\sqrt[3]{3^{4x-3}} = \sqrt[6]{3^{x^2+5}}$.

Both Apollo missions had three crew members. What Apollo missions did make the first manned flight to the Moon and the first man landing on the Moon?

8. The number of all people who have ever walked on the Moon is the smallest sum of Pythagorean triples, which are consecutive terms of an arithmetic sequence. How many people have walked on the Moon?

9. It is difficult to imagine our life without a telephone. *This telephone has too many shortcomings to be seriously considered as a practical form of communication. The device is inherently of no value to us* according to a Western Union internal memo. The memo was prepared in the year such as the numbers formed by the first two digits of the year and its last two digits are the third and thirteen terms of an arithmetic sequence with 90 as the sum of the first five terms. Moreover, the first term and the difference of the sequence are the same. What year the prediction about the telephone was made?

10. Darryl F. Zanuck (1902–1979) was the head of 20th Century Fox. This Hollywood legend who earned three Oscars and a star on the Walk of Fame disbelieved in emerging technologies and predicted that *television won't be able to hold on to any market it captures after the first six months. People will soon get tired of staring at a plywood box every night.*

What year did Darryl Zanuck make this prediction if the sum of the first two digit and the sum of the last two digits of the year are the same? In addition, the difference between the second and first digits is the cube of the difference between the fourth (digits) and third digits. None of the digits in the year is repeated.

Answers

1. 1829
2. 1869, 1969
3. D
4. $5000
5. years
6. 1969, 1936
7. Apollo-8, Apollo-11
8. 12

9. 1878

10. 1946

Solutions

1. Let a, b, c, and d be the first, second, third, and fourth digits of the year. Then $a + b = d$ and $a + b + c + d = dc + 2$ lead to $c + 2d = dc + 2$. Hence, $(c - 2)(d - 1) = 0$, which is possible if either $c = 2$ or $d = 1$, or both $c = 2$ and $d = 1$. The digit d cannot be 1, because then from $a + b = d$ follows that a or b is 1, which does not satisfy the problem (*all digits should be different*). Therefore, $c = 2$. Adding $a + b = d$ and $a + d = b + 2$ leads to $a = 1$ and $b + 1 = d$. Remembering the statement that *the second digit is not less than four times the third digit*, we get that $b = 8$ and $d = 9$. The year is 1829.

   Van Buren wrote his letter in 1829.

2. Applying the binomial expansion (see Appendix I-AF), we rewrite $(x + y)^n$ as

$$C_n^0 x^n y^0 + \ldots + C_n^{\frac{n-1}{2}} x^{n-\frac{n-1}{2}} y^{\frac{n-1}{2}} + C_n^{\frac{n-1}{2}+1} x^{n-\left(\frac{n-1}{2}+1\right)} y^{\frac{n-1}{2}+1} + \ldots + C_n^n x^0 y^n.$$

   Because n is odd, the polynomial has an even number of terms. Using the property $C_n^{\frac{n-1}{2}} = C_n^{\frac{n-1}{2}+1}$ we get $\dfrac{C_n^{\frac{n-1}{2}} x^{n-\frac{n-1}{2}} y^{\frac{n-1}{2}}}{C_n^{\frac{n-1}{2}+1} x^{n-\left(\frac{n-1}{2}+1\right)} y^{\frac{n-1}{2}+1}} = \dfrac{89}{21}$, or $x/y = 89/21$. Because 89 and 21 are co-primes (see Appendix I-PF) then $xy = 1869$.

   The First Transcontinental Railway was open in 1869. The first manned mission landed on the Moon in 1969.

3. Let x mi be the length of the railway built by the Central Pacific Railroad Company of California, then $\frac{22}{115}x$ mi is a part of the rail line built by the Western Pacific Railroad Company and $x - 5\left(\frac{22}{115}x\right) = 30$. Solving the last equation, we obtain $x = 690$ mi. The portion of the rail line built by the Union Pacific Railroad Company is $\frac{217}{138}x$ mi and the total length is $\frac{22}{115} \cdot 690 + 690 + \frac{217}{138} \cdot 690 = 1907$.

   The First Transcontinental Railway is 1907 miles long.

4. Let $\$x$ be the initial amount, then $\$2500x$ is the selling amount and $\$2500x - \$x = \$12{,}495{,}000$, that is, $x = \$5000$.

   Mr. Rackham invested $5000.

5. Let us consider two functions: $y = x^2 - 4x + 9 = (x - 2)^2 + 5$ and $y = 5\cos(x - 2)$. The minimum point of $y = (x - 2)^2 + 5$, which is a parabola that opens up, is (2, 5). The function $y = 5\cos(x - 2)$ reaches its maximum at points $(2 + 2\pi n, 5)$ where n is any integer. Thus, the only intersection point of these two functions is (2, 5) which leads to the solution $x = 2$.

Just 2 years after their prediction of impossibility to fly within 50 years, Wright brothers flew.

6. Only digits 0, 1, 6, 8, and 9 can be written upside down. They will become 0, 1, 9, 8, and 6 correspondingly. The units digit 2 of 4992 can be obtained if a number ended with 9 is subtracted from a number ended with 1 or a number ended with 8 is subtracted from a number ended with 0. The second option does not work because then the first digit of the year would be 0. Hence, the first and last digits of the year are 1 and 9. Looking at 9 and 4992 we can conclude that 9 and 6 are the only options for middle digits of the year.

Apollo left for the Moon in 1969.

7. Raising both sides of $\sqrt[3]{3^{4x-3}} = \sqrt[6]{3^{x^2+5}}$ to the second power, we obtain $(3^{4x-3})^2 = 3^{x^2+5}$, from which follows $8x - 6 = x^2 + 5 \Rightarrow x^2 - 8x + 11 = 0$. The sum of all solutions to the last quadratic equation and $\sqrt[3]{3^{4x-3}} = \sqrt[6]{3^{x^2+5}}$ is 8 and their product is 11.

Apollo-8 made the first manned flight to the Moon on December 21–27, 1968. Apollo-11 had the first man landing on the Moon on July 16–24, 1969.

8. Let a and d be the first term and difference of an arithmetic sequence. Following the Pythagorean theorem, we can write $a^2 + (a + d)^2 = (a + 2d)^2 \Rightarrow 3d^2 + 2ad - n^2 = 0 \Rightarrow d = -a, a/3$. The solution $d = -a$ does not satisfy the problem and $d = a/3$ produces integers if a is divisible by 3: (3, 4, 5), (6, 8, 10), (9, 12, 15), ... , $(3n, 4n, 5n)$, ... The smallest sum is 12.

12 American astronauts waked on the Moon.

9. Let a and b be the numbers formed by the first and last two digits of the year and d be the first term and the difference of the arithmetic sequence. Substituting $n = 5$ and $a_1 = d$ to $S_n = \frac{n(2a_1 + (n-1)d)}{2}$, we get $S_5 = \frac{5(2d + (5-1)d)}{2} = 90 \Rightarrow d = 6$ and $a_1 = 6$. Then the third term of the arithmetic sequence is 18, and its 13th term is 78 (see Appendix I-SS).

The Western Union memo of telephone worthless was made in 1878.

10. Let a, b, c, and d be the first, second, third, and fourth digits of the year, then $b - a = (d - c)^3$. Both sides are one-digit numbers. Only the cubes of three numbers 0, 1, and 2 are one-digit numbers. The first option of 0 leads to the repetition of digits and so does the second option (see $a + b = d + c$). Thus, only 2 is left. Hence, $d - c = 2$ and $b - a = 8$. The statement $b = 8 + a$ leads only to $b = 9$ and $a = 1$. Then, $d = 6$ and $c = 4$. The year is 1946.

Mr. Darryl Zanuck predicted a decline of television in 1946.

V.4 UNESCO Sites and High-Tech Companies

The United Nations or UN was established in San Francisco to promote international cooperation. The name for this intergovernmental organization was suggested by the 32nd President of the United States Franklin D. Roosevelt. One of its agencies, the United Nations Educational, Scientific and Cultural Organization or UNESCO, was founded on 4 November 1946 with headquarter in Paris. UNESCO aims to keep peace by enhancing international educational and cultural connections. Each country is honored to have sites recognized by UNESCO.

To increase a peaceful cooperation further, the USA hosts various cultural, scientific, and sport events. The USA promotes and supports technological development and high-tech companies that grow fast.

1. The sum of all digits of the year when the UN was founded in San Francisco, California, is the number formed by the first two digits of that year. The product of even-position digits of the year is the number formed by its last two digits. The sum of the last two digits is the second digit of the year. What year was the UN founded?

   Furthermore, the product of two middle digits of the year is the square of the number of languages used at UN forums. The number of the UN quarters (one of them is in New York) is the smallest composite factor of that product. How many languages are used at UN forums? How many UN quarters are in the UN?

2. Independence Hall in Pennsylvania, Mesa Verde National Park in Colorado, San Antonio Missions in Texas, and Statue of Liberty in New York are among US cultural sites recognized by UNESCO.

   How many UNESCO cultural sites are in the USA if this number is the largest number in a set of Pythagorean triples formed by three consecutive even numbers?

3. Carlsbad Caverns National Park in New Mexico, Everglades National Park in Florida, Grand Canyon National Park in Arizona, Hawaii Volcanoes National Park in Hawaii, Yosemite National Park in California, and Yellowstone National Park in Wyoming, Montana, and Idaho are among the US natural sites recognized by UNESCO.

How many UNESCO natural sites are in the USA if this number can be formed by three consecutive positive integers in two ways: as a sum of squares of two larger numbers without the square of the smallest one and as a sum of a cube and a square of the middle number?

4. Independence National Historical Park in Philadelphia keeps the memory of the American Revolution and the nation's founding history. Its Independence Hall, where the Declaration of Independence and the United States Constitution were signed, is listed as UNESCO World Heritage Site.

Philadelphia is the largest city of the Commonwealth of Pennsylvania and Harrisburg is its capital.

One cyclist left Harrisburg and another cyclist left the Independence National Historical Park. What would be the distance between them 1 hour before they met? Use the hints below to solve the problem. How many hints do you need?

A. The distance between Harrisburg and Independence National Historical Park is 125 mi via Schuylkill River Trail.

B. The Independence National Historical Park was established on July 4, 1956.

C. The second cyclist left the Independence National Historical Park 2 hours and 10 minutes after the first cyclist left Harrisburg.

D. The cyclists biked toward each other at 15 mph and 18 mph.

5. Grand Canyon National Park, Arizona, became the 15th national park in the United States in 1919 and a UNESCO World Heritage Site in 1979. The Grand Canyon is one of the *Wonders of the World*.

The spectacular North Rim of the Grand Canyon, is

$(1^2 + 3^2 + 5^2 + 7^2 + ...+43^2 + 45^2) - (2^2 + 4^2 + 6^2 + 8^2 + ...+ 42^2 + 44^2)$ feet above its South Rim. How much is the North Rim higher than the South Rim?

6. The USA hosted Olympic Games three times. Los Angeles was the first city to have Olympic games in 1984. The number of years, which is the sum of all solutions to

$$6 - (6 - x)^6 = 6^{|x-6|},$$

passed before Atlanta welcomed the Olympic games, and then Salt Lake City greeted the Olympic Games in a half of this time period. What years did Atlanta and Salt Lake City host Olympic Games?

7.–9. *It is impossible to imagine our everyday life without valuable services and products offered by Google, YouTube, Wikipedia, and other high-tech companies that have grown fast.*

7. The American technology company Google was founded by Ph.D. students of Stanford University, Larry Page and Sergey Brin, as their research project. The headquarter of this multinational giant is in Mountain View, California.

   What year was Google founded if 333, 666, and 999 are among factors of that year?

8. Libraries, books, and other information sources have been used since ancient times. Necessity for an electronic source has increased with rapid technological developments. The first known proposal for an online encyclopedia was made by Rick Gates in 1993. Jimmy Wales and Larry Sanger started Wikipedia on 15 January.

   A pile of beans was divided and put into four bags. The first bag had 7/23 of all beans from the pile, the second one had 6/29, and the third bag had 1/3 of all beans from the pile. The remaining 311 beans were put into the fourth bag. The number of beans is the same as the year Wikipedia was founded. When was Wikipedia founded?

9. The American video-sharing website YouTube headquartered in San Bruno, California, was brought to life by Chad Hurley, Steve Chen, and Jawed Karim. Google bought this site in November, 2006 for $1.65 billion.

   What year was YouTube founded if this year has d as its first (thousands) digit, $a - b$ as its second (hundreds) one, $d - 2$ as its tens, and c as its ones digit, where positive a, b, c, and d can be found from

$$\frac{a + b\sqrt{c}}{d} = \sqrt{1 + \sqrt{1 + \sqrt{1 + \sqrt{1 + \ldots}}}} \; ?$$

10. Built by Howe Technologies, the PAV1 Badger is the smallest all-terrain armored vehicle listed among 50 extreme Guinness World Records. It is powerful enough to break down doors but small enough to fit in an elevator.

How wide is the armored vehicle if its width is x feet x inches where x satisfies the following equation

$$7^{x-3} = 13^{3-x}?$$

11.–12. *US scientists, artists, and politicians have been internationally recognized and received prestigious awards and prices. Functions, units of measurement, buildings, and cities have been named after them.*

11. The Nobel Prize is an international award for outstanding achievements in the fields of physics, chemistry, medicine, literature, economic science, and for peace. Since 1901 the prize ceremony has been provided on December 10 annually, except for 49 years mostly during World War I (1914–1918) and World War II (1939–1945).

How many exceptional American scientists, inventors, and politicians received this prestigious award between 1901 and 2015 if the digits of this number include all values of x that are not in the domain of the function

$$y = \frac{x(x^2 - 12x + 35)}{x^2 - 12x + 35}$$

and their differences, arranged in the increasing order and written side to side to form the number?

12. Paul Howard Douglas was born on March 26, 1892. He was an American politician and economist. The well-known Cobb-Douglas function was named after him and his collaborator. The function has been widely used in economics.

Dr. Douglas spent one quarter of his life and 8 years to grow, study, and earn his PhD in economics. Then he spent one quarter of his life teaching at universities and providing government service at the Chicago City Council. Dr. Douglas enlisted in the United States Marine Corps as a private. After serving there for 1/21 of his life, he retired as a Lieutenant Colonel with two Purple heart and a Bronze Star. Then he came back to a politics campaigning for the Senate for 1/7 of his previous time in politics (teaching and government service) and serving as a US Senator from Illinois for 6 times his campaign time. His retirement lasted a half of the time he served as the US Senate. How many years did Paul Howard Douglas live?

Answers

1. 1945, 4 quarters, 6 languages
2. 10
3. 12
4. 33 mi, D
5. 1035 ft
6. 1996, 2002
7. 1998
8. 2001
9. 2005
10. 3 ft 3 in
11. 257
12. 84

Solutions

1. Let $\overline{abcd}$ be the year. Presenting statements of the problem as mathematical equations, we obtain the system of three equations in four variables $\begin{cases} a + b + c + d = 10a + b \\ bd = 10c + d \\ c + d = b \end{cases}$. Subtracting the third equation from the first, we get $9a = b$, from which follows $a = 1$, $b = 9$, because a and b are digits. Then the second equation becomes $5c = 4d$, which leaves 5 as the only choice for d and 4 for c.

The product of two middle digits is 36 and its square root is 6. The smallest composite factor of 36 is 4.

The UN was founded in 1945 in San Francisco. It has 4 quarters and use 6 official languages.

2. Let m be the largest even integer in a set of Pythagorean triples, then $m - 2$ and $m - 4$ are the other two even integers that form the set of Pythagorean triples. It is obvious that $m \geq 6$. Then from

$$(m - 4)^2 + (m - 2)^2 = m^2$$

follows $m^2 - 12m + 20 = 0$, with $m = 2$, $m = 10$ as its solution. Only $m = 10 \geq 6$ satisfies the problem.
The US has ten UNESCO cultural sites.

3. Let m be the middle integer, then $(m + 1)^2 + m^2 - (m - 1)^2 = m^3 + m^2$, or $m^3 - 4m = 0$ that leads to $m = 2$, $m = -2$, and $m = 0$. Only $m = 2$ satisfies the problem. Then three consequent numbers are 1, 2, 3 and

$$(3)^2 + 2^2 - (1)^2 = 2^3 + 2^2 = 12.$$

The USA has 12 UNESCO natural sites.

4. Two cyclists biked toward each other. They would be $(15 + 18)$ mph times 1 hour before they met. Only Hint D is used.

5. Let us rewrite $(1^2 + 3^2 + 5^2 + 7^2 + ...+43^2 + 45^2) - (2^2 + 4^2 + 6^2 + 8^2 + ... + 42^2 + 44^2)$ as $1^2 + (3^2 - 2^2) + (5^2 - 4^2) + ... +(43^2 - 42^2) + (45^2 - 44^2)$. Using $a^2 - b^2 = (a - b)(a + b)$, the last statement becomes $1 + 5 + 9 + ... +85 + 89$. It is the arithmetic series (see Appendix I-SS) with $a = 1$, $a_{23} = 89$, $d = 4$, $n = 23$. The sum is $\frac{1 + 89}{2}23 = 1035$.

The North Rim of the Grand Canyon National Park is 1035 ft above the South Rim.

6. Let us use the substitution $t = x - 6$ and consider two functions $y = 6 - t^6$ and $z = 6^{|t|}$. The two functions are continuous, monotonic, even, and symmetric about the y-axis. The function $y = 6 - t^6$ opens downward from 6 to $-\infty$. It increases if $t < 0$ and decreases if $t > 0$. The function $z = 6^{|t|}$ opens upward from 1 to ∞. It

increases if $t > 0$ and decreases if $t < 0$. Hence, the graphs of the two functions intersect at two points with $-t^*$ and t^* as their x-coordinates. Then x is $-t^* + 6$ or $t^* + 6$ and the sum of both solutions to $6 - (6 - x)^6 = 6^{|x-6|}$ is $(-t^* + 6) + (t^* + 6) = 12$.

The Olympic Games were in Atlanta in 1996 and in Salt Lake City in 2002.

7. Prime factorization of 333, 666, 999 gives $333 = 3^2 \cdot 37$, $666 = 2 \cdot 3^2 \cdot 37$, $999 = 3^3 \cdot 37$. Hence the year should be divided by at least $2 \cdot 3^3 \cdot 37$, which is 1998.

Google was founded in 1998.

8. Let x be the year when Wikipedia was founded and the number of beans in the bag. From $\frac{7}{23}x + \frac{6}{29}x + \frac{1}{3}x + 311 = x$ follows $x = 2001$.

Another way to solve the problem is to check the denominators of all fractions because the year should be divisible by these numbers. Indeed, $23 \cdot 29 \cdot 3 = 2001$.
Wikipedia was founded in 2001.

9. Let φ be $\sqrt{1 + \sqrt{1 + \sqrt{1 + \sqrt{1 + \ldots}}}}$, then $\varphi = \sqrt{1 + \varphi}$ leads to the quadratic equation $\varphi^2 - \varphi - 1 = 0$ that has a solution $\frac{1 + 1 \cdot \sqrt{5}}{2}$. Hence, $a = 1, b = 1, c = 5, d = 2$, and the year is 2005.

YouTube was founded in 2005.
Remark: $\varphi = \frac{1 + \sqrt{5}}{2}$ is the famous golden ratio.

10. The equality $7^{x-3} = 13^{3-x}$ makes sense only if $x - 3 = 0$ or $x = 3$.
The PAV1 Badger is 3 ft x 3 in.

11. The Domain of the function $y = \frac{x(x^2 - 12x + 35)}{x^2 - 12x + 35}$ consists of all real numbers except for 5 and 7. Their difference is 2. Thus, the number is 257.

257 US scientists, inventors, and politicians were awarded the Nobel Prize between 1901 and 2015.

12. Let x years represent Douglas' life, then $\frac{1}{4}x + 8$ years stand for growing, studying, and earning PhD, $\frac{1}{4}x$ years represent time of teaching at universities and providing government service at the Chicago City Council, $\frac{1}{21}x$ years represent serving time in United States Marine Corps, $\frac{1}{7} \cdot \frac{1}{4}x$ years are for campaigning for the Senate,

$6 \cdot \frac{1}{7} \cdot \frac{1}{4} x$ years are serving as a US Senator from Illinois, $\frac{1}{2} \cdot 6 \cdot \frac{1}{7} \cdot \frac{1}{4} x$ years give retirement time. Hence,

$$\left(\frac{1}{4} x + 8\right) + \left(\frac{1}{4} x\right) + \left(\frac{1}{21} x\right) + \left(\frac{1}{7} \cdot \frac{1}{4} x\right) + \left(6 \cdot \frac{1}{7} \cdot \frac{1}{4} x\right)$$
$$+ \left(\frac{1}{2} \cdot 6 \cdot \frac{1}{7} \cdot \frac{1}{4} x\right) = x \Rightarrow \frac{2}{21} x = 8 \Rightarrow x = 84.$$

$1892 + 84 = 1976.$

Paul Howard Douglas (March 26, 1892–September 24, 1976).

VI

The United States in Arts

DOI: 10.1201/9781003229889-6

VI.1 The American Story in Art

Monuments, sculptures, paintings, and poetry tell a story illustrate the remarkable past of a nation or country.

1. The Liberty Bell is an iconic symbol of American independence. It arrived in Philadelphia in August 1752 with a timeless message, *Proclaim Liberty throughout all the Land unto all the Inhabitants thereof.* The circumference of a circular base of the Liberty Bell is 12 ft. Find its base area.

   A. $\frac{36}{\pi}$ sq. ft

   B. 36 sq. ft

   C. 36π sq. ft

   D. 72 sq. ft

   E. 144 sq. ft

2. The 2013 Perry's Victory and International Peace Memorial, Ohio, honors those who fought in the Battle of Lake Erie during the War of 1812. Commodore Oliver Hazard Perry led his fleet to the victory in this naval battle. The memorial is listed on the National Register of Historic Places. The Memorial Doric column is the *tallest* monument in the USA and is among the most massive Doric columns in the world.

   The height of the Doric column in feet is a three-digit even number. The sum of its first (hundreds) and the last (ones) digits is the middle (tens) digit, and their product is one more than the middle digit. How tall is the Doric?

3. The World War II Memorial in Washington D.C. honors 16 million Americans who served in the military or supported the troops. It was open to the public on April 29, 2004 and became a part of the National Park System later that year. The Memorial granite

columns symbolize unity among American states as of 1945, federal territories, and the District of Columbia.

How tall is each granite column of the World War II Memorial if its height is the solution to $(\log_x 107)^{\sqrt{43+128x-3x^2}} = x^{\sqrt[4]{x^2-41x-86}}$?

4. In their paintings, American illustrators and sculptors Frederic S. Remington and Charles Marion Russell show the life of cowboys and American Indians and landscapes of old American West.

   Remington was born 3 years before Russell but died 17 years earlier. Russell lived 10 years more in the 19th century than in 20th. Who of these great artists lived longer and for how many years? What are years of their birth and death?

5.–10. *Washington Crossing the Delaware. After the loss of New York City and several months of defeat during the American Revolutionary War, troops led by George Washington crossed the Delaware River on the night of December 25–26 and attacked the Hessian forces in Trenton, New Jersey. Despite that only one out of three planned river crossings was successful, this victory raised the spirit of the American colonists.*

This historic attack is remembered by generations. Communities of Washington Crossing in Pennsylvania and New Jersey have been named after it. Paintings by Thomas Sully, Emmanuel Gottlieb Leutze, and Edward Persy Moran depict this historical event. The painting "Washington Crossing the Delaware" by the German American artist Emmanuel Gottlieb Leutze inspired an American lexicographer David Shulman to write the sonnet "Washington Crossing the Delaware".

5. The event, painting, and sonnet of *Washington Crossing the Delaware* appeared in three consecutive centuries. The years of the event and painting are related to as 592 to 617. The last two digits of the years of the event and painting, and of the painting and sonnet are related to as 76 to 51 and 17 to 12 correspondingly. What years are associated with the historical event, painting, and sonnet?

   Note: The last two digits of the years are mentioned in the order they appear in a year.

6. Artists sometimes exaggerate and implement some inaccuracies to highlight an event. Indeed, the boat in the Leutze's painting *Washington Crossing the Delaware* is too small to carry people shown in the painting, but this emphasizes the struggle of the rowing soldiers. The *Stars and Stripes*, not the first national *Grand Union Flag* official at that time, was shown in the painting.

   The *Stars and Stripes* was adopted on June 14 in the number of years after Washington's crossing in 1776. The number of years is the minimum value of the function $f(x) = |x - 6| - |x - 7| + |x - 8|$. When was the *Stars and Stripes* adopted? What are critical points of the function? Is the function continuous? Does the function have a maximum point? Sketch the function.

7. The painting *Washington Crossing the Delaware* by Leutze is remarkable for its artistic composition. The diversity of people in the boat represents the union of the American colonies: a man in a Scottish bonnet, a man in a Native American garb, a man of an African descent, western riflemen, and two farmers.

   How many presidents of the United States are in the painting *Washington Crossing the Delaware* by Leutze if this number is the largest value of the function $f(x) = 8 \cdot (-0.5)^n$ when n increases from 1 to ∞. Can you name the presidents?

8.–10. *Inspired by the painting "Washington Crossing the Delaware" by Leutze, American lexicographer and cryptographer David Shulman (1912–2004) wrote a 14-line rhyming sonnet "Washington Crossing the Delaware", in which every line is an anagram of the title.*

Note: Anagram is a phrase or word formed by rearranging the letters of another phrase or word. For instance, each line of the last two lines of the sonnet:

> "George can't lose war with's hands in;
> He's astern – so go alight, crew, and win!"
> has the same 29 letters as its title "Washington Crossing the Delaware". Check it!

8. Imagine that all letters of the 14 lines of the sonnet *Washington Crossing the Delaware* are mixed and put to a box. What letter has the highest chance to be selected? What is this chance? From which box is the probability of selecting the letter greater, from this box or from the box that has letters only from one line of the sonnet?

9. Let all letters of the 14 lines of the sonnet *Washington Crossing the Delaware* be mixed and put to a box.

   If two letters are selected at random, what is the chance that these two letters are *w* if

   a. A selected letter is put back to the box

   b. A selected letter is kept

   From which box is the probability of selecting this letter greater: from this box or from the second box that has letters only from one line of the sonnet?

10. Imagine that all letters of the title of the sonnet *Washington Crossing the Delaware* are mixed up and put in a box. What is chance that the word *GEORGE* can be formed from six letters randomly selected from the box?

11.–12. *The Bunker Hill Monument was erected between 1825 and 1843 in Charlestown, Massachusetts, at Breed's Hill, the site of the first major battle of the American Revolution fought on June 17, 1775.*

11. How tall is The Bunker Hill Monument if its height in feet is a three-digit number with a as its ones, b as its tens, $\left|\frac{c-d}{ab}\right|$ as its hundreds, and $a\sqrt{b}\cdot(\sqrt{c}+\sqrt{d}) = \frac{\sqrt{5-\sqrt{21}}\cdot(7\sqrt{7}-3\sqrt{3})}{11\sqrt{7}-16\sqrt{3}}$?

12. The Bunker Hill Monument does not have elevators. It is worth climbing the steps to the pinnacle to enjoy a wonderful view of the Boston area.

    The first prime numbers are 2, 3, 5, 7, and 11. The square and cube of the only even prime number are 4 and 8, respectively. It seems that all these numbers can be used to describe the number of steps at the Bunker Hill Monument. Indeed, the number of steps is the smallest number divisible by 2, 3, and 7. It has the remainder of 4 when divided by 5 and 8 when divided by 11. How many steps lead to the top of the Bunker Hill Monument?

Answers

1. A
2. 352 ft
3. 43 ft
4. Russell by 14 years; no
5. 1776, 1851, 1936
6. 1777, {(6, 1), (7, 2), (8, 1)}, yes, no
7. 2
8. a, s, n, e, 3/29, the same
9. 4/841, 2/435, the same, the first box
10. 1/21,375,900
11. 221 ft
12. 294

Solutions

1. The radius of a circle can be expressed via its circumference $C = 2\pi r$ as $r = \frac{C}{2\pi}$. Substituting the last formula to the area of a circle $A = \pi r^2$ and simplifying the obtained expression, we get $A = \pi r^2 = \pi\left(\frac{C}{2\pi}\right)^2 = \frac{144}{4\pi} = \frac{36}{\pi}$.

The base area of the Liberty Bell is $\frac{36}{\pi}$ sq. ft. The correct choice is A.

2. Let a, b, and c be three digits (hundreds, tens, and ones, correspondingly) of a three-digit number that represent the height, $a \geq 1$, c is even. Then the problem can be modeled by the system of two equations in three variables $\begin{cases} a + c = b \\ ac = b + 1 \end{cases}$. The system does not have any positive solution if $a = 1$ or $c = 0$. Hence, $a > 1$ and $c \neq 0$. Then the product $ac \geq 12$ and b is not a digit if $c \geq 6$. Two even digits, 2 and 4, are left for c.

If $c = 2$ then $a = 3$ and $b = 5$. There are no integer values for a and b if $c = 4$. Thus, the only choice is 352.

The 2013 Perry's Victory and International Peace Memorial is 352 ft tall.

3. Since the expression under the square root should be nonnegative, the base of a logarithmic function should be greater than 0 but not 1, and the base of an exponential function should be greater than 0, the domain is $\begin{cases} 43 + 128x - 3x^2 \geq 0 \\ x^2 - 41x - 86 \geq 0 \\ x > 0;\ x \neq 1 \end{cases} \Rightarrow \begin{cases} -(x - 43)(3x + 1) \geq 0 \\ (x - 43)(x + 2) \geq 0 \\ x > 0;\ x \neq 1 \end{cases} \Rightarrow x = 43$

(see Appendix I-EL). Hence, 43 is the only number in the domain.

Let us check whether 43 satisfies the equation. Substituting 43 for x leads to $(\log_{43} 107)^0 = 43^0 \Rightarrow 1 = 1$, which justifies that 43 is the solution to the equation.

The height of each column of the World War II Memorial in Washington is 43 ft.

4. $B_{Russell} - B_{Remington} = 3$; $D_{Russell} - D_{Remington} = 17$. Thus, $17 - 3 = 14$. Russel lived 14 years longer than Remington. It is not enough information to answer the second question.

Remark: Remington (1861–1909), Russell (1864–1926).

5. The last two digits of the years when the event occurred and painting was presented are related to as 76 and 51. If doubled, they will become three-digit numbers. Thus, 76 and 51 are the last two digits of the years. Since $\frac{17}{12} = \frac{51}{36}$ (to obtain 51 as in the previous ratio), the last two digits of the year when the sonnet was written lead to 36. Then $\frac{592}{617} = \frac{1776}{1851}$ (to get two last digits of 76), and the years of the event and painting are 1776 and 1851. Finally, 1936 is the year when the sonnet was written as the only year with two last digits of 36 after the year 1851.

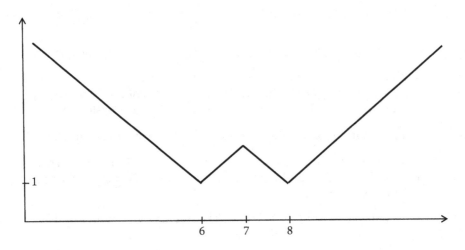

FIGURE VI.1

6. The function $f(x) = |x - 6| - |x - 7| + |x - 8|$ is continuous (see Figure VI.1). The critical points are $(6, 1)$, $(7, 2)$, and $(8, 1)$. The minimum value is 1. To sketch the graph, let us consider the intervals $(-\infty, 6) \cup (6, 7) \cup (7, 8) \cup (8, \infty)$.

On $(-\infty, 6)$, the function becomes $f(x) = -(x - 6) + (x - 7) - (x - 8) = -x + 7$, on $(6, 7)$, it becomes $f(x) = (x - 6) + (x - 7) - (x - 8) = x - 5$, on $(7, 8)$, it becomes $f(x) = (x - 6) - (x - 7) - (x - 8) = -x + 9$ on $(8, \infty)$, it becomes $f(x) = (x - 6) - (x - 7) + (x - 8) = x - 7$.

7. The term $(-0.5)^n = -.5$ if $n = 1$, $(-0.5)^n = 0.25$ if $n \geq 2$, has a decreasing absolute value of the result as n increases. Thus, $-0.5 \leq (-0.5)^n \leq 0.25$ and $f_{max} = 8 \cdot .25 = 2$.

The US Presidents George Washington and James Monroe are depicted in the painting. The man standing next to George Washington and holding the flag is Lieutenant James Monroe, the fifth President of the United States, served between 1817 and 1825. Monroe was the last president who was a Founding Father of the United States.

8. There are $29 \cdot 14$ letters in the sonnet, the letters a, s, n, e appear three times in one line. Thus, the probability of selection any of these letters is $(3 \cdot 14)/(29 \cdot 14)$. If one line of the sonnet is selected, then the probability of selection any of these letters is $3/29$, which is the same.

9. There are $29 \cdot 14$ letters in the sonnet, the letter w appears twice in one line, that is, $2 \cdot 14$ in all lines. Thus, the probability of selection

it at the first time is $(2 \cdot 14)/(29 \cdot 14) = 2/29$. If a selected letter put back, then the probability of selection w again is $2/29$ and the chance of selecting w twice is $(2/29) \cdot (2/29)$. The probability is the same if the selection is made from the second box that has letters only from one line of the sonnet.

If a selected letter is not put back, then the probability of selection w again is $2/29 \cdot (2 \cdot 14 - 1)/(29 \cdot 14 - 1) = 2/435$, but from the second box is $2/29 \cdot 1/28 = 2/812$, which is smaller.

10. The probability of selection two g out of two, two e out of two, one o out of two, and one r out of two letters from 29 letters of the title is $(2/29) \cdot (1/28) \cdot (2/27) \cdot (1/26) \cdot (2/25) \cdot (2/24)$.

11. The formulas of the square and cube of differences (see Appendix I-AF) are applied several times. Indeed, let us simplify $\sqrt{5 - \sqrt{21}}$ and $7\sqrt{7} - 3\sqrt{3}$ as

$$\sqrt{5 - \sqrt{21}} = \sqrt{\frac{2 \cdot 5 - 2\sqrt{21}}{2}} = \sqrt{\frac{(\sqrt{7})^2 - 2\sqrt{7 \cdot 3} + (\sqrt{3})^2}{2}}$$

$$= \sqrt{\frac{(\sqrt{7} - \sqrt{3})^2}{2}} = \frac{\sqrt{7} - \sqrt{3}}{\sqrt{2}};$$

$$7\sqrt{7} - 3\sqrt{3} = (\sqrt{7})^3 - (\sqrt{3})^3 = (\sqrt{7} - \sqrt{3})((\sqrt{7})^2 + \sqrt{21}$$
$$+ (\sqrt{3})^2) = (\sqrt{7} - \sqrt{3})(10 + \sqrt{21}),$$

and, similarly,

$$11\sqrt{7} - 16\sqrt{3} = (10 + \sqrt{21})(2\sqrt{7} - 3\sqrt{3}) = (10 + $$
$$\sqrt{21})\frac{(7\sqrt{7} - 21\sqrt{3} + 9\sqrt{7} - 3\sqrt{3})}{8} = (10 + \sqrt{21})\frac{(\sqrt{7} - \sqrt{3})^3}{8}.$$

Substituting these expressions to the original one, we will obtain

$$\frac{\sqrt{5 - \sqrt{21}} \cdot (7\sqrt{7} - 3\sqrt{3})}{11\sqrt{7} - 16\sqrt{3}} = \frac{8(\sqrt{7} - \sqrt{3})(\sqrt{7} - \sqrt{3})(10 + \sqrt{21})}{\sqrt{2} \cdot (10 + \sqrt{21})(\sqrt{7} - \sqrt{3})^3} = \frac{8}{\sqrt{2} \cdot (\sqrt{7} - \sqrt{3})}$$
$$= \frac{8 \cdot \sqrt{2} \cdot (\sqrt{7} + \sqrt{3})}{\sqrt{2} \cdot \sqrt{2} \cdot (\sqrt{7} - \sqrt{3}) \cdot (\sqrt{7} + \sqrt{3})} = \frac{8 \cdot \sqrt{2} \cdot (\sqrt{7} + \sqrt{3})}{8} = 1 \cdot \sqrt{2} \cdot (\sqrt{7} + \sqrt{3}).$$

Thus, $a = 1$, $b = 2$, $c = 7$ or $c = 3$, $d = 3$ or $d = 7$, and $|c - d| = 4$.

The Bunker Hill Monument is 221 ft tall.

12. The number of steps is divisible by 2, 3, and 7, which are coprimes, then it should be divisible by 42 or can be presented as $42n$, where $n \in N$. Because the number has the remainder of 4 when

divided by 5, then it can be presented as $5k + 4$, $k \in N$. Therefore, $42n = 5k + 4 \Rightarrow k = \frac{42n - 4}{5}$. Since k is a natural number, then only numbers ending with 2 or 7 are suitable for n. Hence, the number of steps $42n$ can be 84, 294, 504, ... The smallest number from this set that has the remainder of 8 when divided by 11 is 294.

294 steps lead to the top of the Bunker Hill Monument.

VI.2 American Writers

American writers are known for their witty, clever, and thought-provoking works. Their stories, poems, and novels have been translated into many languages and enjoyed throughout the entire world – sometimes even more popular in other countries than in America. Among American authors, there are few better known than Mark Twain for his humor, Edgar Allan Poe for his macabre, and O'Henry for his unexpected endings.

1. Well-known American writer Samuel Langhorne Clemens tried many jobs – a printer, a typesetter, a miner, and a riverboat pilot. Most of all, though, he enjoyed writing. His humorous stories received international attention and are still popular today. Do you know him? Have you read any of his stories or books? No? Then look at other hints: Samuel L. Clemens is also famous for his wise quotes, including:

   • *The secret of getting ahead is getting started.*
   • *Thousands of geniuses live and die undiscovered – either by themselves or by others.*
   • *Don't let schooling interfere with your education.*

   Mr. Clemens signed his work as Josh, Thomas Jefferson Snodgrass, and Mark Twain. *Mark Twain* became his most favorite penname that reminded him of his years working on Mississippi riverboats and the cry *mark twain* that meant *the water [according to two marks] is 12 feet deep and it is safe to pass.*

   Now, what books and short stories by Samuel L. Clemens (Mark Twain) have you read?

   Mark Twain was born in Florida, Missouri, on November 30, 1835. The Halley's Comet visited Earth at the year of his birth, and Twain predicted that he would "go out with it". Sadly, his prediction was correct – Mark Twain died from a heart attack on April 21, 1910, in Redding, Connecticut, shortly after Halley's Comet visited Earth again.

How many days after the Halley's Comet visited Earth was Mark Twain born and how many days after its second visit did he die, if these numbers are the denominator and numerator, respectively, of the sum of the series

$$\frac{2}{2\cdot4\cdot6} + \frac{2}{4\cdot6\cdot8} + \frac{2}{6\cdot8\cdot10} + ... + \frac{2}{2n\cdot(2n+2)\cdot(2n+4)} + ...$$

presented as an irreducible fraction?

2. The poems *The Raven* and *Annabel Lee* and stories *The Tell-Tale Heart* and *The Fall of the House of Usher* by Edgar Allan Poe are among the greatest gems of American literature. Edgar Allan Poe was one of the earliest American short story and detective fiction writers. He became more popular in Europe than in the USA and was the first famous American writer that tried to earn a living through writing.

Edgar Allan Poe lived in the 19th century. He was born to a family of two actors on January 19 in Boston and died on October 7 in Baltimore. The year of Poe's death is the square of the number that is 3 more than the number of years he lived. What year was Edgar Allan Poe born in? When did he die?

3.–11. *William Sidney Porter is an internationally recognized master of a short story with twisted plots and surprised endings. Have you read his stories? No? Then read further. Mr. Porter used different pennames such as S. H. Peters, James L. Bliss, T. B. Dowd, Howard Clark, and ...O. Henry. The movie O'Henry's Full House, 1952, is based on his stories, The Clarion Call, The Last Leaf, The Ransom of Red Chief, and The Gift of the Magi. His wise quotes are still popular today:*

- *Write what you like; there is no other rule.*
- *The true adventurer goes forth aimless and uncalculating to meet and greet unknown fate.*
- *It'll be a great place if they ever finish it.*

The US and Soviet Postal Services have issued stamps commemorating the 100th and 150th anniversaries of his birth.

3. O'Henry was born on September 11 in North Carolina. What year was he born in? Use the following hints to answer the question:

- The second digit of the year is the sum of the third and fourth digit
- Triple second digit without four times the third digit is the difference between 10 and five times the fourth digit

- The third digit tripled is the last digit and twice the second digit

4. Mr. Porter moved to Texas, primarily for health reasons, in a year with the last digit as a prime number. The third digit of the year is the last digit cubed. The second digit is a multiple of the last digit not less than the third digit. At least two digits of the year are different. What year did Mr. Porter move to Texas?

5. One of Porter's jobs in Texas was a teller with the First National Bank in Austin. He was falsely accused, found guilty of banking fraud, and was sentenced to 5 years in an Ohio prison.

   The prison number assigned to Porter in the Ohio State Penitentiary was a five-digit number. The first digit was odd. The third and the fourth digits were greater than the first digit and were both its multiples. If the last digit (ones) was divided by the first digit, then the quotient and remainder would be the same. The difference between the fourth and the third digits was the second digit. The sum of all five digits was the smallest number that had the remainder 1 after division by 3, the remainder 3 after division by 4, and the remainder 4 after division by 5. What was his prison number?

6. Three years later Mr. Porter left the Ohio State Penitentiary with a dozen short stories signed as O'Henry to hide his identity. He moved to New York City and published his stories that made him an internationally known America's short story writer.

   The number of his stories rejected by journals is a magic number that does not change any number if written to its left, though creates a miracle if written to the right of the number. How many stories by O'Henry were rejected?

7. O'Henry died in New York. Each of the first three digits of the year of his death are perfect squares, which are not all the same. The second digit can be presented as the difference of the prime number formed by the first two digits of the year and the number made from its last two digits. When did O'Henry die?

8. The fortune in dollars left by O'Henry after he died was
   $$\frac{\left(\sqrt{\frac{3}{5}}+\sqrt{\frac{4}{5}}\right)\left(1\frac{1}{6}\cdot\frac{4}{7}+.1\right)}{2\frac{1}{3}-1\frac{2}{3}}\cdot\left(\sqrt{\frac{4}{5}}-\sqrt{\frac{3}{5}}\right).$$ Find his fortune without a calculator.

9. O'Henry's 400 short stories are known for their wit and twist endings. The average number of typed pages in his stories satisfies the equation $6x^3 - 31x^2 - 105x = 0$. This number was his hat size as well. What is the average number of pages in O'Henry's stories?

10. The number of stories that O'Henry wrote about Texas is a two-digit number with the first digit (tens) squared being the second digit (ones) and the difference between the digits being the first digit doubled. How many stories about Texas did O'Henry write?

11. O'Henry's work had been translated to several foreign languages by 1923. This number can be presented by a set of four consecutive primes as the sum of two prime numbers or the product of two others. The sum of the four primes is the smallest. In how many foreign languages could people enjoy reading O'Henry's stories in 1923?

Answers

1. 16 days, 1 day
2. 1809, 1849
3. 1862
4. 1882
5. 30664
6. 0
7. 1910
8. 23 cents
9. 7½
10. 39
11. 10

Solutions

1. Let us consider the n-th term of the series, $\dfrac{2}{2n \cdot (2n + 2) \cdot (2n + 4)}$. Factoring 2 from the second and third factors in the denominator, we get $\dfrac{2}{2n \cdot (2n + 2) \cdot (2n + 4)} = \dfrac{2}{8n \cdot (n + 1) \cdot (n + 2)}$. Following Partial-Fraction Decomposition Technique, we can present the last fraction as a combination of simple fractions as $\dfrac{1}{4n \cdot (n + 1) \cdot (n + 2)} = \dfrac{1}{4}\left(\dfrac{A}{n} + \dfrac{B}{n + 1} + \dfrac{C}{n + 2}\right)$, or $\dfrac{1}{4n \cdot (n + 1) \cdot (n + 2)} = \dfrac{1}{4}\dfrac{A(n + 1)(n + 2) + Bn(n + 2) + Cn(n + 1)}{n(n + 1)(n + 2)}$. Two fractions with the same denominators are equal if their numerators are the same, i.e., $1 = A(n + 1)(n + 2) + Bn(n + 2) + Cn(n + 1)$.

The last equation is supposed to be true for any value of n. So, we can pick any values of n, substitute to the equation, and find the unknown coefficients A, B, and C. Indeed, let us take $n = -1$, then the equation becomes $1 = A \cdot 0 \cdot 1 + B \cdot (-1) \cdot 1 + C \cdot (-1) \cdot 0$ or $B = -1$.

If $n = 0$, then $1 = A \cdot 1 \cdot 2 + B \cdot 0 \cdot 1 + C \cdot 0 \cdot 1$ or $A = 1/2$.

Finally, if $n = -2$, then $1 = A \cdot (-1) \cdot 0 + B \cdot (-2) \cdot 0 + C \cdot (-2) \cdot (-1)$, or $C = 1/2$.

Using $\frac{1}{4n \cdot (n+1) \cdot (n+2)} = \frac{1}{4}\left(\frac{1/2}{n} - \frac{1}{n+1} + \frac{1/2}{n+2}\right)$, we can rewrite our series

$$\frac{2}{2 \cdot 4 \cdot 6} + \frac{2}{4 \cdot 6 \cdot 8} + \frac{2}{6 \cdot 8 \cdot 10} + \ldots + \frac{2}{2n \cdot (2n+2) \cdot (2n+4)} + \ldots \text{ as}$$

$$\frac{2}{8}\left(\frac{1}{1 \cdot 2 \cdot 3} + \frac{1}{2 \cdot 3 \cdot 4} + \frac{2}{3 \cdot 4 \cdot 5} + \ldots + \frac{2}{n \cdot (n+1) \cdot (n+2)} + \ldots\right)$$

$$= \frac{1}{4}\left(\frac{1/2}{1} - \frac{1}{2} + \frac{1/2}{3} + \frac{1/2}{2} - \frac{1}{3} + \frac{1/2}{4} + \frac{1/2}{3} - \frac{1}{4} + \frac{1/2}{5} + \ldots + \frac{1/2}{n} - \frac{1}{n+1} + \frac{1/2}{n+2} + \ldots\right).$$

Note that $\frac{1/2}{3}$, $-\frac{1}{3}$, and $\frac{1/2}{3}$; $\frac{1/2}{4}$, $-\frac{1}{4}$, and $\frac{1/2}{4}$; and the next consecutive triple terms are cancelled out, leaving $\frac{1}{4}\left(\frac{1/2}{1} - \frac{1}{2} + \frac{1/2}{2}\right) = \frac{1}{16}$.

Mark Twain was born 16 days and died 1 day after the Halley's Comet visited Earth.

Note. There are other ways to find the coefficients A, B, and C. For instance, they can be found by equating the coefficients of polynomials in the numerators

$$0 \cdot n^2 + 0 \cdot n + 1 = An^2 + 3An + 2A + Bn^2 + 2Bn + Cn^2 + Cn,$$

and solving the system of three equations

$$
\begin{array}{ll}
n^2 \quad A + B + C = 0 & A = 1/2 \\
n \quad 3A + 2B + C = 0. \text{ The solution to the system is } & B = -1 \ . \\
n^0 \quad 2A = 1 & C = 1/2
\end{array}
$$

2. 1849 is between $1600 = 40^2$ and $2500 = 50^2$. Using a trial-and-error method, we can find that 43 is the only number such that $43^2 = 1849$ falls between 1800 and 1899 (19th century).

Edgar Allan Poe was born in 1809 and died in 1849 when he was 40.

3. Obviously, the first digit of the year is 1. Let a, b, and c be the second, third, and forth digits of the year, then

$$
\begin{cases}
a = b + c \\
3a - 4b = 10 - 5c \\
2a + c = 3b
\end{cases}
\Rightarrow
\begin{cases}
a - b - c = 0 \\
3a - 4b + 5c = 10. \\
2a - 3b + c = 0
\end{cases}
$$

Adding the first and third equations we get $3a - 4b = 0$. Then the second equation becomes $5c = 10$ or $c = 2$. Substituting $c = 2$ to the system, we obtain $\begin{cases} a - b - 2 = 0 \\ 3a - 4b + 10 = 10 \\ 2a - 3b + 2 = 0 \end{cases}$. Multiplying the first equation by -2 and adding it to the third equation, we get $-b + 6 = 0$, from which follows $b = 6$. Substituting $b = 6$ and $c = 2$ to any equation, we get $a = 8$.

Hence, O'Henry was born in 1862.

4. There are three one-digit numbers, 0, 1, and 8, that are another number cubed. The numbers 0 and 1 do not satisfy other conditions of the problem. Then 8 is left and the last digit is 2. The third digit is 8. The second digit is a factor of the last digit (which can be 2, 4, 6, 8) not less than the third digit (only 8 is left).

Mr. Porter moved to Texas 1882.

5. 1st way. The sum S of all digits is 19. Indeed, the sum S can be presented as $5n + 4$ or $4m + 3$, or $3l + 1$, where n, m, and l are natural numbers. Equating the first and the last expressions, we get $n = \frac{3(l-1)}{5}$. Because 3 and 5 are co-primes, $l - 1$ should be divisible by 5, that is, l can be 6, 11, 16, ... , and the sum S can be 19, 34, 49, ... Checking the lowest sum 19 we can see, that 19 divided by 3 has the reminder of 1, and 19 divided by 4 has the remainder of 3.

2nd way: numbers that have the remainder 4 after division by 5 are 9, 14, 19, 24, 29, ... Checking these numbers for the remainder 3 after division by 4, we will find that the lowest number is 19. Finally, 19 divided by 3 has the reminder of 1 and 19 divided by 4 has the remainder of 3.

To solve the problem further, let us consider odd one-digit numbers for the first digit of the O'Henry's prison number: 1, 3, 5, 7, 9. The numbers 5 or greater than 5 cannot be taken because their multiples are two-digit numbers. The number 1 cannot be chosen, because any number is divisible by 1 without a remainder. Then the only option for the first digit is 3.

The choices for the third and the fourth digits are 6 and 9.

The fifth digit can be 4 ($= 3 \cdot 1 + 1$) or 8 ($= 3 \cdot 2 + 2$). The number 8 does not work because the difference 19 (the sum) $- 8$ (the fifth digit) $- 3$ (the first digit) is 8 that cannot represent the sum of the second, third, and fourth digits. Then the fifth digit is 4 and $19 - 4 - 3 = 12$. Hence, the third and fourth digits can be only 6 and the second digit is 0.

The O'Henry's prison number was 30,664.

6. The digit 0 makes no difference if written to the left but changes the number if written to the right. For example, writing 0 to the left of 5

does not change its value, and writing 0 to the right of 5 makes 50. O'Henry claimed that he did not have any stories rejected. To get such a result, he simply used a new envelope, put a new stamp, and re-sent the same story to another journal.

7. One-digit perfect squares 1, 4, 9 can stand for the first three digits. From the numbers formed by the first two digits of the year, namely 11, 14, 19, 91, 41, the number 14 is not prime; 11 does not work either because then all digits are the same. Finally, 91 and 41 are meaningless. Thus, only 19 is left. Finally, $19 - 9 = 10$, and the year is 1910. O' Henry died in 1910.

8. Using the basic formulas $\frac{a}{b}c = \frac{ac}{b}$ and $(a - b)(a + b) = a^2 - b^2$ (see Appendix I-AF), we get

$$\frac{\left(\sqrt{\frac{3}{5}} + \sqrt{\frac{4}{5}}\right)\left(1\frac{1}{6} \cdot \frac{4}{7} + .1\right)}{2\frac{1}{3} - 1\frac{2}{3}} \cdot \left(\sqrt{\frac{4}{5}} - \sqrt{\frac{3}{5}}\right) =$$

$$= \frac{\left(\sqrt{\frac{3}{5}} + \sqrt{\frac{4}{5}}\right) \cdot \left(\sqrt{\frac{4}{5}} - \sqrt{\frac{3}{5}}\right) \cdot \left(\frac{7}{6} \cdot \frac{4}{7} + .1\right)}{2\frac{1}{3} - 1\frac{2}{3}} = \frac{\left(\frac{4}{5} - \frac{3}{5}\right)\left(\frac{2}{3} + \frac{1}{10}\right)}{\frac{2}{3}} =$$

$$= \frac{\left(\frac{4-3}{5}\right)\left(\frac{20+3}{30}\right)}{\frac{2}{3}} = \frac{\left(\frac{1}{5}\right)\left(\frac{23}{30}\right)}{\frac{2}{3}} = \frac{1 \cdot 23 \cdot 3}{2 \cdot 5 \cdot 30} = \frac{23}{100} = .23.$$

As an alcoholic, O'Henry died virtually penniless.

9. The equation $6x^3 - 31x^2 - 105x = x(6x^2 - 31x - 105) = 0$ can be rewritten as $x = 0$ and $6x^2 - 31x - 105 = 0$. Applying the quadratic formula to the last equation $x = \frac{31 \pm \sqrt{31^2 - 4 \cdot 6 \cdot (-105)}}{2 \cdot 6} = \frac{31 \pm \sqrt{31^2 - 4 \cdot 6 \cdot (-105)}}{2 \cdot 6} = \frac{31 \pm \sqrt{3481}}{12} = \frac{31 \pm 59}{12}$, we can find the solution to the original equation as $\{0, -7/3, 15/2\}$.

The average number of typed pages in O'Henry's stories is 7½.

10. 1st way. Let a be the first digit and b be the second, then
$\begin{cases} a^2 = b \\ b - a = 2a \end{cases}$. Substituting b from the first equation to the second, we get $a^2 - a = 2a \Rightarrow a(a - 3) = 0 \Rightarrow a = 3, b = 9$.

2nd way: Let us consider different options:

First Digit	Second Digit	Difference
1	1	0
2	4	4
3	9	$6 = 2 \cdot 3$

O'Henry wrote 39 stories about Texas.

11. Let us consider the first four consecutive primes, namely, 2, 3, 5, 7. Checking different combinations we get: $2 \cdot 5 = 10$ and $3 + 7 = 10$. O'Henry's stories were translated into 10 languages by 1923.

VI.3 American Women in Arts

American women have greatly contributed to the success and prosperity of the nation. They have won prestigious awards in almost all areas of the arts and sciences. However, recognition has not come easily to them. American women have had to fight for freedoms and equal rights and wages. Stories about nine American females famous for their literature and artistic achievements are presented below, but the list of remarkable women can be continued.

1.–2. *The list of American women to be awarded the prestigious Nobel Prize began in 1931 with sociologist Jane Addams, who co-won the Nobel Peace Prize for her service as a social worker and political activist. American women Pearl S. Buck and Toni Morrison received the Nobel Price for their accomplishments in literature.*

1. American writer and novelist Pearl Sydenstricker Buck was born in Hillsboro, West Virginia. Being a daughter of missionaries, she spent her childhood in China and later wrote about Chinese and Asian cultures. Her novel, *The Good Earth*, was the bestselling fiction book in the USA and won the Pulitzer Prize in 1932. Pearl Buck became the first American woman to be awarded the Nobel Prize for Literature for *her rich and truly epic descriptions of peasant life in China and for her biographical masterpieces.* Pearl Buck was also a prominent advocate for the rights of women and other minority groups. She was honored with the US commemorative stamp from the Great Americans series. Buck died from lung cancer at the age of 80 in Danby, Vermont, but her legacy remains in her novels and her wise quotes:

   • *To know how to do something well is to enjoy it*

   • *If you want to understand today, you have to search yesterday*

   • *I don't wait for moods. You accomplish nothing if you do that. Your mind must know it has got to get down to work*

   • *The young do not know enough to be prudent, and therefore they attempt the impossible – and achieve it, generation after generation*

   The number formed with the last two digits of the year in the 20th century in which Pearl Buck was awarded the Nobel Prize are the year first two digits doubled. When did Pearl Buck receive her Nobel Prize?

Note: All digits are taken in the order they appear in the year.

2. The first African American female to be awarded the Nobel Prize in Literature was Toni Morrison, an American novelist, editor, and professor at Princeton University. She also won a Pulitzer Prize and the American Book Award in 1988 for her book, *Beloved*, which was later adapted into a movie. She has been honored with other prestigious awards and medals, including the Presidential Medal of Freedom, presented to her by President Barak Obama in 2012.

Toni Morrison was born Chloe Ardelia Wofford on February 18, 1931. She received the Nobel Prize at the age that

a. Was either twice the number formed by the first two digits (taken in any order) of the year she was born or twice the number formed by the last two digits of that year

b. Was neither a multiple of 4 nor a perfect square

c. Had the sum of all digits that is either less than 10 or greater than 11

d. Was a multiple of neither 11 nor 13

When did Toni Morrison receive the Nobel Prize? Are all four hints necessary to solve the problem?

3. The first African American female poet to be published in America, Phillis Wheatley, was born in West Africa. She was sold into slavery at the age of seven and brought to North America, where the Wheatley family of Boston purchased her. The Wheatleys educated her and encouraged her to write poetry. At the age of 20, Phillis Wheatley published her first book, *Poems on Various Subjects, Religious and Moral*, which brought her fame in both England and the American colonies. President George Washington admired her work. Phillis married after she was emancipated. After her husband was imprisoned for debt, Phillis fell into poverty and died of illness.

All four digits but the first one of the year of Phillis's birth are different prime numbers placed in a decreasing order. The sum of the first two digits of the year she was born is equal the sum of its last two digits. Furthermore, the sum of the first two digits of the year of Phillis's death is equal to its third digit and twice the fourth digit. When did Phillis Wheatly live?

4. African American journalist, newspaper editor, sociologist, and feminist Ida Bell Wells-Barnett was born on July 16 in Holly Springs, Mississippi. She became one of the leaders in the Civil Rights Movement and one of the founders of the National

Association for the Advancement of Colored People. During her European tours, Mrs. Wells campaigned for justice. *The Red Record: Tabulated Statistics and Alleged Causes of Lynching in the United States, 1895,* and *Southern Horrors: Lynch Law in All Its Phases, 1892,* are among her most famous books. A musical drama, *Constant Star* (2006), by Tazewell Thompson, tells a story of her life.

The third digit of the year when Ida Bell Wells-Barnett was born is the square root of the product of the last digit and the number composed of the first two digits. The difference between the second and the fourth digits of the year is the third digit. When was Ida Wells born?

5. American writer and civil rights activist Maya Angelou published seven autobiographies and several books of essays and poetry. Her first book, *I Know Why the Caged Bird Sings,* was published in 1969 and brought her international recognition. She was also an actor and producer, as well as a spokesperson for black people and a defender of black culture. Her encouraging quotes appeal to each heart:

- *If you don't like something, change it. If you can't change it, change your attitude*
- *Nothing will work unless you do*
- *We may encounter many defeats, but we must not be defeated*

Maya Angelou received dozens of awards and more than 50 honorary degrees. She recited her poem, *On the Pulse of Morning,* at President Bill Clinton's inauguration in 1993. President Barack Obama awarded her the Presidential Medal of Freedom in 2011.

Maya Angelou was born on April 4 in St. Louis, Missouri, and passed away on May 28 in Winston-Salem, North Carolina. The years of her birth and death are the seventh and ninth terms of one arithmetic sequence and the fifteenth and fifty-eighth terms of another arithmetic sequence. The common differences of the two arithmetic sequences are prime numbers. The sum of the first terms of the two sequences is 3570. When was Maya Angelou born? When did she pass away?

6.–7. *The famous American author and journalist Margaret Munnerlyn Mitchell was born in Atlanta, Georgia, into a wealthy and politically prominent family. Her most recognizable novel "Gone with the Wind" won several awards. Margaret Mitchell was married twice. Her second marriage was to John Marsh, who had been the best man at her first wedding.*

6. Margaret Mitchell was struck by a speeding drunk driver and died 5 days later. If she had lived one more year, then the greatest common divisor of the years of her birth and death would have

been 50, and the years would be related to as 38 to 39. When was Margaret Mitchell born? When did she pass away?

7. For her novel *Gone with the Wind*, Margaret Mitchell won the National book award for the Most Distinguished Novel in the only year in the 20th century presented as a square number. This novel was also awarded the Pulitzer Prize for Fiction one year later. When did Margaret Mitchell win these two prestigious awards?

8. Alice Walker (born February 9, 1944) is an American novelist, poet, and activist. She wrote many wonderful novels, such as *Meridian* (1976), *The Third Life of Grange Copeland*, and *The Color Purple* (1982). Walker won the National Book Award for Hardcover Fiction and the Pulitzer Prize for Fiction for the latter novel. As a feminist, she coined the term *womanist* meaning a *black feminist or feminist of col*or in 1983.

   Walker was the youngest among her seven siblings. When she was eight, one of her brothers fired a BB gun and injured her right eye, making her permanently blind in that eye. She did not give up her dreams, but instead increased her hours spent on reading and writing.

   At what age did Alice Walker wrote her novel, *The Color Purple*, if the age x is the fifth term of the following sequence 98, 66, 50, 42, x, 36, 35, ...?

9. American painter and printmaker Mary Stevenson Cassatt was born in Allegheny City, Pennsylvania. Against her parents' wishes, she followed her dream to become a painter and moved to Paris. Her work was exhibited in the Paris Salon with the Impressionists and is the jewel of any museum and private collection. Her paintings depict the social and private lives of women and their bonds with kids.

   Mary Stevenson Cassatt was born on May 22, 1844 and passed away in Paris at the age equaled the first term of the following sequence: _, 54, 36, 28, 30, 42, ... When did she die?

10. American artist Georgia Totto O'Keeffe was born on November 15 in Wisconsin to a family of dairy farmers. When she was a little girl, O'Keeffe decided to become an artist, though she abandoned and returned to her dream several times throughout her life. She spent a lot of time in New Mexico before moving there perma-nently in 1949. Georgia O'Keeffe is best remembered for her paintings of enlarged flowers, New York skyscrapers, and New Mexico landscapes. Her paintings are among treasures of any museum. One of O'Keeffe's paintings, *Jimson Weed/White Flower No. 1*, 1932, sold for $44,405,000 in 2004, over three times the previous world auction record for any female artist. The US Postal Service issued a 32-cent stamp honoring O'Keeffe in 1996 and a stamp featuring O'Keeffe's *Black Mesa Landscape, New Mexico/Out*

Back of Marie's II, 1930 as part of their Modern Art in America series in 2013. O'Keeffe had a long and productive life.

How old was O'Keeffe when she died, if her age can be presented as the product of x and y, which bring the minimum value to the function $y = 1887x^2 - 3774x + 1986$? Have you noticed anything special about the equation?

Answers

1. 1938
2. 1993, no, B
3. 1753–1784
4. 1862
5. 1928–2014
6. 1900–1949
7. 1936, 1937
8. 1982
9. 1926
10. 99

Solutions

1. Since Pearl Buck lived in the 20th century, the first two digits are 19, which lead to $19 \cdot 2 = 38$ and the year of 1938.
   Ms. Pearl Buck was awarded the Nobel Prize in Literature in 1938.

2. Four numbers can be formed from the first and last two digits: 19, 91, 31, and 13. These numbers twice produce 38, 182, 62, and 26. The second option 182 is not realistic. The other three, 38, 62, and 26, pass Hint B (therefore, it is not necessary), but only 62 and 26 satisfy Hint C. The number 26 is divisible by 13 (Hint D) and should be removed. Thus, only 62 is left, which leads to $1931 + 62 = 1993$.
   Ms. Tony Morrison received the Nobel Prize in Literature in 1993.

3. Conditions that all digits but the first one of the year of Wheatley's birth are different primes (2, 3, 5, 7) placed in a decreasing order, and that the sum of the first two digits of the year is equal to the sum of the last two digits lead to the only choice of $1 + 7 = 5 + 3$. Thus, the year of her birth is 1753. The first two digits of the year of her dearth can be 17 or 18. Since the third digit, which is the sum of the first two digits, is twice the fourth digit, only 17 is left and the last two digits are 84.

The first African American author of a published book of poetry, Phillis Wheatley Peters, was born in 1753 and died in 1784.

4. 1st option: Considering possible numbers formed by the first two digits, 17, 18, and 19, we can see that only $18 = 3^2 \cdot 2$ requires multiplication by a one-digit number 2 or 8 to end up with a perfect square as their product. The second option 8 does not work because the square root of $18 \cdot 8$ is a two-digit number. Then the first two digits of the year are 1 and 8, the third digit is 2, and the fourth digit is $8 - 2 = 6$. The year is 1862.

Two other options 17 and 19 need to be multiplied by two-digit numbers to produce a perfect square which is unsuitable for the tens digit.

Other possible options that satisfy the first part of the problem are the years of 1641, 1263, etc., but they do not satisfy the second condition that the difference between the second and the fourth digits is the third digit.

2nd option: Obviously, the first digit of the year is 1, though the problem can be solved without this assumption. Let a, b, and c be the second, third, and forth digits of the year of birth, then $(10 + a)$ is the number formed by the first two digits and $\begin{cases} b = \sqrt{c(10 + a)} \\ a - c = b \end{cases}$.

This system of two equations can be solved in different ways, two of which are described below.

1st way. Combining two equations of the system, we get $a^2 - 2ac + c^2 = 10c + ac$ that leads to the quadratic equation (see Appendix I-QE) $c^2 - c(10 + 3a) + a^2 = 0$ with respect to one-digit integer c. The solution $c = \frac{(10 + 3a) \pm \sqrt{(10 + 3a)^2 - 4a^2}}{2}$ is integer if $(10 + 3a)^2 - 4a^2 = 5a^2 + 60a + 100$ is a perfect square of a one-digit number a, which can be true for $a = 8$. Then $c = 2$ and $b = 6$.

2nd way. Let us consider two options:

A. Both $(10 + a)$ and c are perfect squares. Then $a = 6$ because 16 is the only a perfect square with the first digit of 1, and c can be 1, 4, or 9. Additionally, $\sqrt{c}$ should be not greater than 2 to keep just one digit for $b = \sqrt{c(10 + a)} = 4\sqrt{c}$. Thus, c could be 1 or 4. Neither of them satisfy the second equation.

B. Neither $(10 + a)$ nor c are perfect squares. Then they can be presented as pr^2 and pq^2 to keep b integer, where p, r, and q are positive integers, and p is not a perfect square. The original system is rewritten in the new notations as $\begin{cases} pr^2 - 10 - pq^2 = b \\ b = prq \end{cases}$.

From its second equation: $0 < pr^2 - 10 - pq^2 \leq 9 \Rightarrow 10 < p$ $(r - q)(r + q) \leq 19$. The product is a composite number. There are 3 composite numbers 12, 16, and 18 between 10 (not inclusive) and 19. Let us consider each of them:

12 – Noticing that $r - q$ and $r + q$ are odd or even at the same time and presenting 12 as the product of its factors we can see that only $p = 4$, $r - q = 1$, and $r + q = 3$ satisfy these requirements, which contradicts that p is not a perfect square.

16 – does not lead to any solutions.
18 – produces $p = 2$, $r - q = 1$, and $r + q = 9$, which lead to $p = 2$, $r = 5$, $q = 4$, and to 1862.

African American journalist and feminist Ida Bell Wells-Barnett was born in 1862.

5. Let a and d, and b and q be the first term and the common difference of the first and second arithmetic sequences, then
$$\begin{cases} a + 6d = b + 14q \\ a + 8d = b + 57q \end{cases} \Rightarrow 2d = 43q.$$ Because d and q are primes, then d = 43, q = 2. Substituting $d = 43$, $q = 2$ to the system leads to $b - a = 230$. Using $a + b = 3570$, we get $a = 1670$, $b = 1900$ and the years of 1928 and 2014.

American writer and civil rights activist Maya Angelou was born in 1928 and died in 2014.

6. Let b and d be years of Margaret Mitchell birth and death. Then $\frac{b}{d+1}$ is reduced to $\frac{38}{39}$, which is valid if both b and $(d + 1)$ are divisible by the same number (their common divisor). Thus, $\frac{b}{d+1} = \frac{38}{39} = \frac{38 \cdot 50}{39 \cdot 50} = \frac{1900}{1950}$, and $d = 1950 - 1 = 1949$.

Margaret Mitchell was born in 1900 and died in 1949.

7. Noticing that $43^2 = 1849$, $44^2 = 1936$, $45^2 = 2025$ we conclude that 1936.
Margaret Mitchell won the Most Distinguished Novel award in 1936.

8. Let us take a look at the differences of the terms of the sequence and at the ratio of the differences:

the sequence	98		66		50		42				x		36		35
the difference		32		16		8				$42 - x$		$x - 36$				1
the ratio				2		2		$8/(42 - x)$						$(x - 36)/1$		

Thus, $x = 38$ and $1944 + 38 = 1982$.

Alice Walker wrote her famous novel *The Color Purple* in 1982.

9. Let us take a look at the differences of the terms of the sequence and at the differences of the differences:

the sequence	_ 54		36		28		30		42
the difference		18		8		− 2		− 12	
the difference			10		10		10		

Thus, the difference between the first and second terms is 28. The first term is 82, which is the number of years artist Cassatt lived. She passed away in $1844 + 82 = 1926$.
Artist Cassatt passed away in 1926.

10. The x- and y-coordinates of the minimum point of the parabola $y = 1887x^2 - 3774x + 1986$ that opens up (the leading coefficient 1887 > 0) are $\left(\frac{3774}{2 \cdot 1887}, 1986 - \frac{3774^2}{4 \cdot 1887} \right)$ or (1, 99).

Artist Georgia Totto O'Keeffe died when she was 99.

Remark. It is worth to mention that the equation coefficients are the years of O'Keeffe's birth and death (1887–1986): $1887x^2 - 2 \cdot 1887x + 1986 = 0$.

VI.4 Awards in Art

Oscar, Emmy, and Tony Awards are among the most prestigious awards for achievements in performing arts. They are recognized all over the word.
1.–2. *Conductor and composer Carl Davis was born on October 28, 1936 in Brooklyn, New York, USA. He moved to the UK in 1961 and was awarded the Most Excellent Order of the British Empire from Her Majesty Queen Elizabeth II. Carl Davis has composed music for more than 100 TV shows, films, and stage works.*

1. How many ballets composed by Carl Davis has been performed at theaters if this number satisfies the equation

$$b^3 - 10b^2 - 20b + 176 = 0?$$

Can you name the ballets composed by Carl Davis?

2. How many stage works by Carl Davis have been performed at theaters if this number satisfies the equation

$$(10 - s)(s^4 - 20) = 5104?$$

Can you name the stage works composed by Carl Davis?

3. Composer Richard Rodgers (1902, New York City – 1979, Pennsylvania) and librettist Oscar Greeley C. Hammerstein II (1895, New York City – 1960, Pennsylvania) started their collaboration in 1943. It turned to be very productive and successful. Their first musical *Oklahoma!* was followed by other musicals and movies for which they received Pulitzer Prizes, Oscars, Emmys, Grammys, among others.

How many musicals did Rodgers and Hammerstein write together if this number is the missing term x in the sequence

$$- 2, -2, 0, 4, x, 18, 28, 40, 54, ...?$$

What is the next term of the sequence? Can you name musicals by Rodgers and Hammerstein? Have you seen any of them?

4. An American composer and lyricist Stephen Joshua Sondheim was born on March 22 in New York. He received more Tony Awards than any other composer. Pulitzer Prize, Academy, Tony and Grammy Awards were given for his lyrics to *West Side Story* and *Gypsy* and music and lyrics to *Company* and *A Little Night Music.*

When was Stephen Sondheim born if the second (hundreds) digit of this year is three times the third (tens) digit, which is three times the first (thousands) digit? The third digit is three more than the last (ones) digit. How many Tony Awards did Stephen Sondheim receive if this number is the difference between the second and first digits of the year of his birth? Have you seen any of his musicals?

5. Since 1950, a winner of the *Oscar* statuette should agree that neither the winner nor their heirs may sell the statuettes without first offering to sell them back to the Academy for

$$\log_2\log_3\log_4\log_5\log_6\log_7\log_8\log_9 9^{8^{7^{6^{5^{4^{3^{2^1}}}}}}}$$

US dollars. The Academy keeps the statuette if the agreement is not reached. How much US dollars does the Academy pay for the Oscar statuette?

6. The Oscar statuette designer, Sachin Smith, used the alloy of 92.5% tin and 7.5% copper before gold-plating it. The company

that casts the *Oscar* statuette has two alloys. One alloy has 96% tin and 4% copper. Another alloy has 88% tin and 12% copper. How much of each alloy is to be taken to make one Oscar statuette that weights 8.5 lb?

7. Sachin Smith who designed the *Oscar* statuette used the alloy of 92.5% tin and 7.5% copper before gold-plated it. A company that cast the Oscar statuette for the award ceremony has the alloy that contains 72 lb more tin than copper. To get the alloy required to cast Oscar statuettes, the company adds copper that weights 20/37 of the weight of the copper in the original alloy. How much copper is added? What is the weight of the obtained alloy? What is the content of the original alloy?

8. Two companies have been assigned to make a certain number of *Emmy Award* statuettes. It takes 48 hours for the first company to complete the order if it works alone and 52 hours for the second company to perform the same job. How much time (in hours and minutes) does it take to complete the task if both companies work together? Is it important to know, that it takes at average five-and-one-half hours to make each Emmy Award statuette?

9. The *Premier Emmy* statuette weights 88 oz. There are three types of alloy of different weight. The total weight of the three alloys is 264 lb. To make the statuette from each alloy, one-fifth from the first alloy is added to the second. Then one-twenty-third from the newly obtained second alloy is added to the third alloy. Finally, after three-twenty-fifth of the third alloy is added to the first alloy, the three alloys become 88 oz each as required. What is the initial weight of each alloy? Can the problem be solved with less information given?

10. The Antoinette Perry Award for Excellence in Theatre or the *Tony Award*, recognizes achievement in Broadway theater, while the *Emmy Award* or Emmy acknowledges excellence in the television industry. The Tony Award medallion was designed by Herman Rosse and the Emmy statuette was created by Lois McManus.

    The difference of squares of the weights in kilograms of the Tony medallion and the Regional Emmy statuette is 3/5, while the difference of the fourth powers of their weights is 339/125. What are the weights of the Tony medallion and the Regional Emmy statuette?

11. Even a good work can be initially rejected. *Chicken Soup for the Writer's Soul* tells a funny story. Several years after the novel *Steps*

won the National Book Award for Fiction in 1969, its author Jerzy Kosinski was permitted to submit the manuscript with a new title and new author's name to several publishing agencies and publishers. They all rejected it, including Random house that had published it in 1968.

The number of years between the National Book Award for Fiction was awarded in 1969 and the novel *Steps'* re-submission and the number of agencies it was submitted to are integers x and y that satisfy the following equation

$$3x^2 - 2xy + 7x - 6y + 38 = 0.$$

Find the year of the manuscript re-submission. How many agencies and publishers rejected it if the number of agencies is one less than the number of publishers?

Answers

1. 4
2. 6
3. 10, 70
4. 1930, 8
5. US$1
6. 4.78 lb, 3.72 lb
7. 2.16 lb, 82.16, 95% tin, 5% copper
8. 24 h and 58 minute, no
9. 95 oz, 73 oz, 96 oz, yes
10. 1.6 kg, 1.4 kg
11. 1977, 13 agencies, 14 publishers

Solutions

1. 1st way. The solution can be found considering all different factors of 176 to find positive integer solutions to $b^3 - 10b^2 - 20b + 176 = 0$.

   2nd way. Only positive integer solutions to the equation $b^3 - 10b^2 - 20b + 176 = 0$ are of our interest. The equation can be rewritten as $(b - 10)(b^2 - 20) = 24$. The right side 24 is presented as the product of $(b - 10)$ and $(b^2 - 20)$, which should be odd or even at the same time. Then, pairs $2 \cdot 6$, $4 \cdot 8$, $(-2) \cdot (-6)$, $(-4) \cdot (-8)$ with the product of 24 are considered.

Furthermore, $(b - 10)$ and $(b^2 - 20)$ should have the same sign, which can be positive if $b > 10$ or negative if $b \leq 4 < \sqrt{20}$. Comparing with options available, b can be 12 or 4. Only 4 is a factor of 176 and can be a solution (see Appendix I-PE). Indeed, by the synthetic division, the equation $b^3 - 10b^2 - 20b + 176 = 0$ can be rewritten as $(b - 4)(b^2 - 6b - 44) = 0$. The last equation has three different solutions: $b = 4$, $b = 3 \pm \sqrt{53}$.

Four ballets by Carl Davis: Lady of the Camellias, Cyrano, Aladdin, Lipizzaner.

2. Let us present the original equation $(10 - s)(s^4 - 20) = 5104$ as $s^5 - 10s^4 - 20s + 5304 = 0$.

Since the sum $10s^4 + 20s^2$ ends with 0, the last digit of $s^5 + 5304$ should be also 0, which is possible if the last digit of s^5 is 6. The last digit of 4 is also possible for a negative s, but only positive integer solutions to the equation are of our interest. Moreover, 6 is a multiple of 5304 and satisfies the equation. Then

$$s^5 - 10s^4 - 20s + 5304 = 0 \Rightarrow (s - 6)(s^4 - 4s^3 - 24s^2 - 144s - 884) = 0$$

Let us justify that 6 is the only integer solution. Rewriting $s^4 - 4s^3 - 24s^2 - 144s - 884 = 0$ as $s^4 = 2^2(s^3 + 6s^2 + 36s - 221)$, we can see that if another integer solution s exists then it is even. Thus, s^4 should be divisible at least by 2^4, which contradicts the fact that the expression $s^3 + 6s^2 + 36s - 221$ is odd when s is even. Thus, the only integer solution of the equation $(10 - s)(s^4 - 20) = 5104$ is 6.

Six stage works by Carl Davis: two musical sets, *The Mermaid* and *Alice in Wonderland*, and four ballets, *Lady of the Camellias, Cyrano, Aladdin, Lipizzaner*.

3. The sequence $-2, -2, 0, 4, x, 18, 28, 40, 54 \ldots$ is neither arithmetic nor geometric sequence. Let us take a look at the difference between terms: $0, 2, 4, x - 4, 18 - x, 10, 12, 14, \ldots$ It is the arithmetic sequence with 0 as the first term and 2 as the common difference. Hence, $x = 10$ and the next term following 54 is 70.

10 musicals by Rodgers and Hammerstein: Oklahoma!, Carousel, Allegro, South Pacific, The King and I, Me and Juliet, Pipe Dream, Flower Drum, Song, The Sound of Music, State Fair

4. 1st way. If the second digit of the year when Stephen Sondheim was born is three times the third one, which is three times the first digit, then the second digit is nine times the first digit. It is

possible if the second digit is 9 and the first one is 1. Then, the second digit is 3 and the last one is 0.

2nd way. Let x be the first digit of the year, then $3x$ is its third digit and $9x$ is the second digit, which is possible if $x = 1$.
Stephen Sondheim was born in 1930. He has received 8 Tony awards.

5. Using properties of logarithms (see Appendix I-EL) $\log_2 \log_3$

$$\log_4 \log_5 \log_6 \log_7 \log_8 \log_9 98^{7^{6^{5^{4^{3^1}}}}} = \log_2 \log_3 \log_4 \log_5 \log_6 \log_7 \log_8$$

$$(8^{7^{6^{5^{4^{3^1}}}}} \log_9 9) = \log_2 \log_3 \log_4 \log_5 \log_6 \log_7 \log_8 8^{7^{6^{5^{4^{3^1}}}}} = \dots =$$
$\log_2 2 = 1$. That gives the value of \$1.

6. Let x lb be the amount of the first alloy to be taken, then $(8.5 - x)$ lb is the amount of the second alloy to be taken. The required amount of tin is $0.96x + 0.88(8.5 - x) = 0.925 \cdot 8.5$, that leads to $x = 4.78$. Thus, 4.78 lb of the first alloy and 3.72 of the second alloy are required to make one Oscar statuette.

If the amount of copper is considered, then $0.04x + 0.12(8.5 - x) = 0.075 \cdot 8.5$ leads to the same result of $x = 4.78$.

7. Let t lb and c lb be the amount of tin and copper in the original alloy that a company has, then $t - c = 72$ and the weight of the original alloy is $t + c = 72 + 2c$. If $20/37$ of the weight of the copper in the original alloy is added, then the weight of the alloy becomes $72 + 2c + 20/37c$ with required 92.5% tin and 7.5% copper. The amount of tin stays the same. That leads to the equation $0.925(72 + 2c + 20/37c) = t$, i.e., $0.925(72 + 2c + 20/37c) = 72 + c$ (if tin is considered) or to $0.075(72 + 2c + 20/37c) = c + 20/37c$ (if copper is considered).

Solving any of these equations, we get: $27c = 108$. The amount of copper and tin in the original alloy are 4 lb and 76 lb, respectively.

The initial alloy consists of $\dfrac{80}{76} \dfrac{100}{x} \Rightarrow$ 95% tin and 5% copper. The weight of the obtained alloy $72 + 2c + 20/37c$ becomes 82.16 lb.

8. Let 1 be the entire work and x be the time needed for two companies working together to complete the assignment. Then

$$\frac{1}{48}x + \frac{1}{52}x = 1 \Rightarrow \frac{25}{624}x = 1 \Rightarrow x = \frac{624}{25} = 24.96h,$$

Transforming .96 hours to minutes we get $\dfrac{96}{x} \dfrac{100}{60} \Rightarrow x = 57.6$ min.

9. Let x oz, y oz, and z oz be the initial weights of each alloy. If one-fifth 1/5 from the first alloy is added to the second alloy, then the weight of the first alloy becomes $\frac{4}{5}x$ and the weight of the second alloy becomes $\frac{1}{5}x + y$ which equals 88 oz after $\frac{1}{23}$ is taken from it or $\frac{22}{23}\left(\frac{1}{5}x + y\right) = 88$. The changes of weights are described in the table below.

	1st Alloy	2nd Alloy	3rd Alloy
At the beginning	x	y	z
After the 1st move	$\frac{4}{5}x$	$\frac{1}{5}x + y$	z
After the 2nd move	$\frac{4}{5}x$	$\frac{22}{23}\left(\frac{1}{5}x + y\right) = 88$	$z + \frac{1}{23}\left(\frac{1}{5}x + y\right)$
After the 3rd move	$\frac{4}{5}x + \frac{3}{25}\left(\frac{1}{23}\left(\frac{1}{5}x + y\right) + z\right) = 88$	$\frac{22}{23}\left(\frac{1}{5}x + y\right) = 88$	$\frac{22}{25}\left(\frac{1}{23}\left(\frac{1}{5}x + y\right) + z\right) = 88$

Based on this information we can write down the system of four linear equations in three variables:

$$
\begin{cases}
\frac{22}{23}\left(\frac{1}{5}x + y\right) = 88 \\
\frac{22}{25}\left(\frac{1}{23}\left(\frac{1}{5}x + y\right) + z\right) = 88 \\
\frac{4}{5}x + \frac{3}{25}\left(\frac{1}{23}\left(\frac{1}{5}x + y\right) + z\right) = 88 \\
x + y + z = 264
\end{cases}
$$

Substituting $\frac{1}{5}x + y = 92$ from the first equation to the second equation and solving it for z, we obtain $z = 96$. Combining these results with the fourth equation we get the system a system of two linear equations in two variables $\begin{cases} \frac{1}{5}x + y = 92 \\ x + y + 96 = 264 \end{cases}$ that leads to $x = 95$, $y = 73$, $z = 96$.

The problem can be solved without one equation or introducing less variables because the resulting system is a system of *four* linear equations in *three* (one less than the number of equations) variables. Indeed, the third equation has not been used while solving the original system of equations.

10. Let t and e be the weight of the Tony medallion and the Emmy statuette, then (see Appendix I-AF) $\begin{cases} t^2 - e^2 = \frac{3}{5} \\ t^4 - e^4 = \frac{339}{125} \end{cases} \Rightarrow$

$\begin{cases} t^2 - e^2 = \frac{3}{5} \\ (t^2 - e^2)(t^2 + e^2) = \frac{339}{125} \end{cases} \Rightarrow \begin{cases} t^2 - e^2 = \frac{3}{5} \\ \frac{3}{5}(t^2 + e^2) = \frac{339}{125} \end{cases} \Rightarrow \begin{cases} t^2 = \frac{64}{25} \\ e^2 = \frac{49}{25} \end{cases} \Rightarrow \begin{cases} t = 1.6 \\ e = 1.4 \end{cases}$.

The weight of the Tony medallion is 1.6 kg (3 1/2lb) and of the Regional Emmy statuette is 1.4 kg (48 oz).

11. Let us rewrite the equation $3x^2 - 2xy + 7x - 6y + 38 = 0 \Leftrightarrow$ $2y(3 + x) = 3x^2 + 7x + 38$

$\Leftrightarrow y = \frac{3x^2 + 7x + 38}{2(3 + x)} = \frac{3x}{2} + \frac{22}{3 + x} - 1$. Because x and y are nonnegative integers then

- Following $\frac{22}{3+x}$, $(3 + x)$ should be a factor of 22, which are 1, 2, 11, and 22. Neither 1 nor 2 work, because then $x = -2$ or -1. The factor 11 leads to $x = 8$. The factor 22 gives $x = 19$.

- From $\frac{3x}{2}$, $3x$ should be even (divisible by 2), then only $x = 8$ is left.

If $x = 8$, then $y = 13$ and they satisfy the equation.

The novel was re-submitted in 8 years, in 1977 to 13 agencies and 14 publishers.

VII

Shopping, Food, and Entertainment in the United States

VII.1 Health Indicators

Obesity increases the risk of heart disease, stroke, high cholesterol, and other health problems. The US federal agencies issue guidelines on defining a healthy weight range. The body mass index BMI, the body surface area BSA, and waist-to-height ratio WHtR are commonly used to identify the weight category: underweight, healthy, overweight, or obese. These measures are good indicators for an average person, though their limitations lead to using various formulas for their calculation.

1. The Body Mass Index $BMI = \frac{W}{H^2}$ and Body Surface Area $BSA = \frac{\sqrt{W \cdot H}}{6}$ of a person are commonly used to estimate the body fat. The person's weight W is in kilograms and the height H is in meters. A 1.69 m tall person lost 9 kg. Find a correct statement from the options below.

   1. BMI has decreased by 25.7 units.
   2. BMI has decreased by 15.21 units.
   3. BMI has decreased by 9 units.
   4. BMI has decreased by 5.33 units.
   5. BMI has decreased by 3.15 units.
   6. Not enough information to compare BMI.
   7. BSA has decreased by 2.535 units.
   8. BSA has decreased by 0.65 units.
   9. BSA has decreased by 0.422 units.
   10. Not enough information to compare BSA.

2. The Body Mass Index $BMI = \frac{W}{H^2}$ is used to estimate the body fat, where the weight W is in kilograms and the height H is in meters. The BMI range of a person in the normal weight category is $18.5 \leq BMI \leq 25$. A person is underweight if $BMI \leq 18.5$, and overweight if $BMI \geq 25$. A 5 ft 3 in person weighs 160 lb. Is this person

DOI: 10.1201/9781003229889-7

underweight or overweight? How many pounds should he/she lose or gain to be in a normal weight category?

3. The Body Mass Index $BMI = \frac{W}{H^2}$ of a person is one of methods to estimate the body fat. The weight W is in kilograms and the height H is in meters. The BMI range of a person in a normal weight category is $18.5 \le BMI \le 25$. What should be the weight of a 5 ft 7 in tall person to stay in a normal category by the BMI standards?

4. The Body Mass Index $BMI = \frac{W}{H^2}$ and the Body Surface Area $BSA = \frac{\sqrt{W \cdot H}}{60}$ are used to estimate the body fat of a person. In both formulas, W is a person's weight in kilograms, but H is their height in meters in the BMI formula and in centimeters in the BSA formula. What is the relation between BSA and BMI for a 1.69 m tall person?

5. The Body Mass Index $BMI = \frac{W}{H^2}$ of a person is used to estimate the body fat. The weight W is measured in kilograms and the height H is in meters. Which formula calculates the BMI if the weight is given in pounds and the height is in inches?

A	$\frac{W\,(lb)}{H^2\,(sq.\,in)} \cdot 0.014$
B	$\frac{W\,(lb)}{H^2\,(sq.\,in)} \cdot 0.06$
C	$\frac{W\,(lb)}{H^2\,(sq.\,in)}$
D	$\frac{W\,(lb)}{H^2\,(sq.\,in)} \cdot 17.9$
E	$\frac{W\,(lb)}{H^2\,(sq.\,in)} \cdot 703$

6. The Body Mass Index BMI is used to determine whether a person is underweight, in normal weight range, or obese. Different formulas for BMI have been suggested. One of them, called BMI *Prime*, is the ratio of actual BMI to the upper limit BMI for a normal weight, which is taken to be 25 in the USA. What is the BMI Prime of a person with BMI of 35 kg/m^2? Interpret this result. What are units of measurement of BMI Prime? Complete the following table:

Category	BMI Range – kg/m^2	BMI Prime
Severely underweight	Less than 16.0	
Underweight	From 16.0 to 18.5	
Normal (healthy weight)	From 18.5 to 25	
Overweight	From 25 to 30	
Obese	Over 30	

7. Various formulas for calculating the body surface area *BSA* have been suggested. A weight-based formula for preteens, proposed by Costeff, is $BSA_{kids} = \frac{4W+7}{W+90}$, where W is a child weight in kilograms. What should be the range of the healthy preteen's weight if the recommended *BSA* range is $1.07 \le BSA_{kids} \le 1.33$?

8. A waist circumference is one of methods for assessing the body fat. The waist-to-height ratio $WHtR = \frac{T}{H}$ of a person measures the distribution of body fat. Here T is the waist circumference, and H is the height. *WHtR* over 0.5 is critical and indicates an increased risk of obesity related diseases. Jessica is 5 ft 6 in tall and her waist circumference is 29 in. What is her *WHtR*? If her measurements are taken in centimeters what will be her *WHtR*? What is the unit of measurement of the *WHtR*? Is Jessica in a risk group?

9. The Body Mass Index *BMI* and waist-to-height ratio *WHtR* are among common measures to estimate the body fat. They are calculated by $BMI = \frac{W}{H^2}$ and $WHtR = \frac{T}{H}$, where W is the weight in kilograms, H is the height meters, and T is the waist circumference in meters. The *BMI* of 25, which is equivalent to *WHtR* of 0.51, is a healthy cutoff value. Find the dependence of the waist circumference on the weight and height for a healthy person.

10. The waist-to-height ratio $WHtR = \frac{T}{H}$ measures the distribution of body fat. In the formula T is the waist circumference, and H is the height. $WHtR = 0.25$ for the Barbie Doll. If Barbie were 5 ft 4 in what would be the measurement of her waist?

Answers

1. 5, 10

2. overweight, 19 lb.

3. [53.5 kg, 72.3 kg] or [118 lb, 160 lb]

4. $BSA = 0.367\sqrt{BMI}$

5. *E*

6. 1.4, 40% over, dimensionless

7. [30.48 kg, 42.21 kg], [67.07 lb, 93.06 lb]

8. 0.43, 0.43, dimensionless, no

9. $T = 0.102\sqrt{W}$, $T = 0.51H$

10. 16 in

Solutions

Hint: Use Appendix I-MU to refresh your knowledge of measurement units.

1. Let L be the weight loss. $BMI_{new} = \frac{W-L}{H^2} \Rightarrow BMI_{new} = BMI - \frac{L}{H^2} \Rightarrow$
$BMI_{new} - BMI = -\frac{L}{H^2}$; which leads to #5.

$BSA_{new} = \frac{\sqrt{(W-L)H}}{6} \Rightarrow BSA_{new}^2 = \frac{(W-L)H}{36} \Rightarrow BSA_{new}^2 = BSA^2 - \frac{LH}{36} \Rightarrow$
$BSA_{new}^2 - BSA^2 = -\frac{LH}{36} \Rightarrow BSA_{new} - BSA = -\frac{LH}{36}/(BSA_{new} + BSA)$.

Thus, either the original weight or BSA is needed, which justifies

the answer #10.

2. $BMI = \frac{W}{H^2} \Rightarrow W = BMI \cdot H^2$ and 5 ft 3 in is 1.6 m. Then,
$18.5 \leq BMI \leq 25 \Rightarrow 18.5 \cdot 1.6^2 \leq W \leq 25 \cdot 1.6^2 \Rightarrow 47.36 \leq W \leq 64$ *kg*, or

in pounds $104.4 \leq W \leq 141.1 lb$. Finally, $160 - 141 = 19$.

The person should lose at least 19 lb but no more than 56 lb.

3. $BMI = \frac{W}{H^2} \Rightarrow W = BMI \cdot H^2$, 5 ft 7 in is 1.7 m, $18.5 \leq BMI \leq 25 \Rightarrow$
$18.5 \cdot 1.7^2 \leq W \leq 25 \cdot 1.7^2$.

The weight should be in the range: [53.46 kg, 72.25 kg] or [117.9 lb, 159.28 lb].

4. $BMI = \frac{W}{H^2} \Rightarrow W = BMI \cdot H^2$. Because 1 m is 100 cm, $BSA = \frac{\sqrt{W \cdot H cm}}{60} =$
$\frac{\sqrt{W \cdot H m}}{6}$, then $BSA = \frac{H\sqrt{BMI \cdot H}}{6}$ or $BSA = \frac{1.69\sqrt{BMI \cdot 1.69}}{6} = 0.367\sqrt{BMI}$.

5. Since 1 lb = 0.45359 kg and 1 in = 0.0254 m, $BMI = \frac{W(kg)}{H^2(m^2)} =$
$\frac{W(lb) \cdot 0.45359(kg/lb)}{H^2(sq.in) \cdot 0.0254^2(m/inc)^2} = \frac{W(lb)}{H^2(sq.in)} \cdot 703$, which is E.

6. The BMI Prime index is dimensionless because it is the ratio between two BMI values. A BMI Prime of 35/25 = 1.4, that is 40% over the recommended upper mass limit.

Category	BMI Range – kg/m²	BMI Prime
Severely underweight	Less than 16.0	Less than 0.64
Underweight	From 16.0 to 18.5	From 0.64 to 0.74
Normal (healthy weight)	From 18.5 to 25	From 0.74 to 1.0
Overweight	From 25 to 30	From 1.0 to 1.2
Obese	Over 30	Over 1.2

7. $BSA_{kids} = \frac{4W+7}{W+90} \Rightarrow W = \frac{90BSA_{kids} - 7}{4 - BSA_{kids}}$;

$1.07 \leq BSA_{kids} \leq 1.33 \Rightarrow 30.48kg \leq W = \frac{90BSA_{kids} - 7}{4 - BSA_{kids}} \leq 42.21kg$.

The healthy weight for preteens is between 30.48 kg and 42.21 kg.

8. Since 5 ft 6 in = 66 in, then $WHtR = \frac{T}{H} = \frac{29}{66} = 0.44 < 0.5$.

$WHtR$ is dimensionless because $WHtR = \frac{Tin}{Hin}$.

9. $BMI = 25 = \frac{W}{H^2} \Rightarrow H = \frac{\sqrt{W}}{5}$, then $WHtR = .51 = \frac{T}{H} = \frac{5T}{\sqrt{W}} \Rightarrow T = 0.102\sqrt{W}.T = 0.51H$.

10. Since 5 ft 4 in is 64 in, then $WHtR = \frac{T}{H} \Rightarrow .25 = \frac{T}{64} \Rightarrow T = 16$.

The Barbie Doll waist would be 16 inches.

VII.2 Food Production

According to the US Department of Agriculture, berries (strawberries, raspberries, blueberries, and cranberries) are the fifth most popular fruits consumed in the nation, after bananas, citrus, apples, and watermelons. Americans also like pastas, pizzas, and hamburgers.

1. Fresh, frozen, and canned strawberries are used in food and drink recipes and cosmetic products. The USA is the leader of the global strawberry market. California is by far the largest producer of strawberries followed by Florida.

   Strawberries contain 92% of moisture per volume. A bulk of 100 lb of strawberries was left to dry. After some moisture evaporated, the strawberry bulk contained 80% of moisture. What was the new weight of the strawberry bulk?

2. Grapes are a natural source of antioxidants and vitamins. A cup of grapes contains just 60 calories and no fat. The top grape-growing US state is California that produces 87% grapes, followed by Washington and New York.

   Grapes contain 90% of moisture. A bulk of grapes was left to dry. After half of moisture evaporated, the bulk became 18 lb lighter. What was the original weight of the bulk?

3. Although cherries are native to Eastern Europe and Asia Minor, they grow well in the USA. Washington, California, and Oregon are its primary producing states. Tart cherries are popular in Michigan.

   A bulk of 100 lb of cherries was left to dry. The bulk became half of its initial weight when the amount of moisture in the bulk went down to 60%. How much moisture is in fresh cherries?

4. Georgia used to be called *The Peach State*, though it is currently ranked fourth in the peach production. Peanuts has become a top crop in Georgia, while California has taken the leading position in growing peach trees. California is also the largest world producer of dried peach.

   Peaches contain 88% of moisture. 100 lb of peaches were left to dry. What was the weight of the dried peaches when they had the equal amount of moisture and flesh content?

5. Illinois and Iowa have competed for the top spot in corn and soybean production for years. However, each state takes the leading spot in growing some other crops. Iowa is a champion in growing green peas, while *The Land of Lincoln* grows more ornamental and eating pumpkins than any other state or about 90% of the US pumpkins.

   One side of a rectangular pumpkin field is increased by 10% and another side is decreased by 10%.

   Compare the new and original areas of the field.

   A. The area stays the same.

   B. The area becomes 99% of the original area.

   C. The area becomes 101% of the original area.

   D. The area becomes 110% of the original area.

6. The flowering plant and edible fruit, watermelon, came to the USA from southern Africa.

   The world's heaviest watermelon was grown by Lloyd Bright of Arkadelphia, Arkansas, in 2005. This record was beat in 2013 when Chris Kent of Sevierville, Tennessee, had grown an 82-lb heavier watermelon. If the Lloyd Bright had a feast and shared his watermelon with his family and friends, then 3 lb would have been left for the last feast day if either 14 lb of this watermelon were eaten each day for several days or 19 lb were eaten each day for a different number of days. Find the weight of two heaviest watermelons. Take the lowest integer values for the numbers of feast days.

7. Many fruits and vegetables have been brought to the USA. Native to Western India, a juicy jackfruit arrived in Hawaii in 1888. Cabbage came from Europe. Its green, purple, or white heads generally weight from 1 to 9 lb.

   The heaviest jackfruit was 76 lb and 22.625 in long. It was grown by George and Margaret Schattauer of Captain Cook, Hawaii, in 2003. The world-heaviest green cabbage was grown by mechanical designer John Evans in Anchorage, Alaska in 1998. It was just

$$\log_{\frac{\pi}{3}}\left(\tan\frac{\pi}{3}\right)\cdot\log_{\frac{\pi}{4}}\left(\tan\frac{\pi}{4}\right)\cdot\log_{\frac{\pi}{5}}\left(\tan\frac{\pi}{5}\right)\cdot\log_{\frac{\pi}{6}}\left(\tan\frac{\pi}{6}\right)\cdot\log_{\frac{\pi}{7}}\left(\tan\frac{\pi}{7}\right)1b$$

heavier than the heaviest jackfruit. What was the weight of heaviest green cabbage?

8.–9. *The largest hamburger on the menu at Mallie's Sports Grill & Bar in Southgate, Michigan, on August 29, 2008, is not only in the Guinness book of records but also among 50 Guinness World Records as well.*

8. The world largest hamburger made in Southgate, Michigan, in 2008, weighed 164.8 lb. How much did it cost, if $5.34 was paid for 1 kg? Round your answer to the nearest integers.

9. What is the weight in kilograms of the world largest hamburger made in Southgate, Michigan, in 2008, if this weight can be presented by solutions to the equation

$$(x - 74)^{x-7} = (x - 74)^{x-4}$$

in a way that the smallest root of the equation stands for the integer part and the largest root shows the decimal part of the weight value?

10. Pizza Fino has served Houston since 1992 offering a "create your own" pizza for the prize presented in the table:

"Create Your Own" Pizza (Diameter)	Small 10"	Medium 12"	Large 14"	X-Large 16"
Cheese pizza	$5.99	$8.99	$9.99	$11.99
Topping (each)	$1.00	$1.25	$1.50	$1.75

Is there a linear relation between the price for cheese pizza and its size (diameter)? What should be the price for medium and large pizzas to have a linear relation between the price for cheese pizza and its size keeping the price of the small and X-large pizzas as advertised? Is there a linear relation between a topping price and a pizza size? Find the size of cheese pizza that has the best value with respect to its size (area).

11. The market for gluten-free food has exploded in recent years. According to the research firm Packaged Facts, the market has grown at the rate of 28% a year since 2008 reaching $4.2 billion in sales in 2012. What was the sales amount in billion dollars

of gluten-free food in 2008? Was the sale in billion dollars of gluten-free food in 2008 the same if it were calculated (a) using the instantaneous rate of 28% since 2008 or (b) if 28% were the annual rate?

Answers

1. 40 lb
2. 40 lb
3. 80%
4. 24 lb
5. B
6. 269 lb, 351 lb
7. 76 lb
8. $399.18
9. 74.75 kg
10. no, $7.99, the same, no, yes, X-Large
11. $1.56 bln at 28% a year, $1.37 bln at the instantaneous rate

Solutions

1. Because of 92% of moisture in strawberries, 100 lb of berries contain just 8 lb of flesh. When berries contain 80% of moisture, 8 lb of flesh take 100% − 80% = 20%, and, thus,

$$\begin{matrix} 8lb & 20\% \\ x & 100\% \end{matrix} \Rightarrow x = \frac{8 \cdot 100}{20} = 40lb.$$

2. Let x lb be the original weight.

Weight	Flesh	Moisture
x lb	0.1x lb	0.9x lb
(x − 18) lb	0.1x lb	0.45x lb

Hence, (x − 18) = 0.1x + 0.45x leads to 40 lb.

3. 1st way: Let x% be the water amount in fresh cherries. Hence, there is x lb of moisture in 100 lb, and (100 − x) lb is the weight of the cherry flesh. When the moisture level drops down to 60%, the cherry flesh takes 40%, and 100 − x = .4·50 ⇒ x = 80.

2nd way: 60% of moisture in the bulk of 50 lb is 30 lb leaving 20 lb for flesh or 80% of moisture initially.

4. 88% of moisture content in 100 lb leaves 12 lb for flesh that after adding the equal amount of moisture gives 24 lb of dried peaches.

5. Let a and b be the sides of a rectangle, then $1.1a \cdot 0.9b = 0.99ab$.

6. 1st way: Let n and k be the numbers of days of the two feasts, then $14n + 3 = 19k + 3$. Thus, $14n = 19k$. Because 14 and 19 are co-primes, then the lowest n and k are 19 and 14 and the weight of the 2005 heaviest watermelon is 269 lb and $269 + 82 = 351$ lb of the 2013-watermelon.

2nd way: Because 14 and 19 are co-primes, then the remainder 3 added to their product will lead to the result: $14 \cdot 19 + 3 = 269$.

7. Since $\tan \frac{\pi}{4} = 1$, and $\log_{\frac{\pi}{4}}\left(\tan \frac{\pi}{4} \right) = \log_{\frac{\pi}{4}} 1 = 0$ (see Appendix I-EL

and TF), then

$$\log_{\frac{\pi}{3}}\left(\tan \frac{\pi}{3} \right) \cdot \log_{\frac{\pi}{4}}\left(\tan \frac{\pi}{4} \right) \cdot \log_{\frac{\pi}{5}}\left(\tan \frac{\pi}{5} \right) \cdot \log_{\frac{\pi}{6}}\left(\tan \frac{\pi}{6} \right) \cdot \log_{\frac{\pi}{7}}\left(\tan \frac{\pi}{7} \right) = 0.$$

8. 164.8 pounds is 74.752 kg; $74.752 \cdot 5.34 = \$399.18$.

9. The bases in both sides are the same, then $x - 7 = x - 4$ that does not have any solution. Hence, the equation is valid only if $x - 74 = 0$ or $x - 74 = 1$, that is $x = 74$ or $x = 75$, because 1 raised to any power is 1 and 0 raised to any positive power is 0.

10. The line that passes through $(10, 5.99)$ and $(16, 11.99)$ is $\frac{y - 5.99}{11.99 - 5.99} = \frac{x - 10}{16 - 10} \Rightarrow y = x - 4.01$.

The price of a medium pizza should have been $y(12) = \$7.99$, (not $\$8.99$ as advertised and, thus, there is no linear regression), but the price of a large pizza should remain the same $y(14) = \$9.99$.

The line that passes through $(10, 1)$ and $(16, 1.75)$ is $y = 8x + 2$ and, checking the topping price and price, we can conclude that there is a linear relation between the price of the topping and its size.

$$A = \pi r^2 \Rightarrow$$

Create Your Own Pizza	Small 10"	Medium 12"	Large 14"	X-Large 16"
Cheese pizza	$5.99	$8.99	$9.99	$11.99
Area sq. inches	78.54	113.10	153.94	201.06
Amount of pizza/$	13.11	12.58	15.40	16.77

11. Let x billion be the sales of gluten-free food in 2008. Then the sales in 2012 became $(1 + .28)^4 x = 4.2 \Rightarrow x = 4.2/(1 + .28)^4 = \$1.564$. In the case of the 28% instantaneous increase, the sales in billion dollars of gluten-free food in 2012 became $xe^{4 \times 0.28} = 4.2 \Rightarrow x = 4.2e^{-4 \times 0.28} = \$1.37$.

The sales of gluten-free food in 2008 was $1.56 billion in the case of the 28% annual rate and $1.37 billion in the case of the instantaneous rate of 28%.

VII.3 Shopping in the USA

Americans love shopping. US department stores, shopping malls, outlets, and boutiques can fit any customer taste.

1. *Lord & Taylor* is the oldest mid- to high-range department store chain and the second oldest retail chain in the USA. It was founded by Samuel Lord and George Washington Taylor in 1826 on Catherine Street in Manhattan, New York, to sell women's apparel. Becoming the president of *Lord & Taylor* in 1945, Dorothy Shaver was the first woman in the USA to head a major retail company.

   A woman bought a dress and paid $58 and a half of what it cost. What was the original price of the dress?

   A. $58

   B. $87

   C. $116

   D. $145

   E. $232

2. *Brooks Brothers* is the oldest US retail chain and one of the nation's oldest continuously operating businesses. On April 7, 1818, Henry Sands Brooks opened H. & D.H. Brooks & Co in New York. In 1870, Brooks' four sons inherited the family business and renamed it *Brooks Brothers*. The merchant is famous for introducing ready-to-wear men suits. It has over 200 stores in the USA and about 100 in other countries.

A suit was bought at *Brooks Brothers* for $100 when it had a 20% discount. What was its original price?

A. $80

B. $120

C. $125

D. $180

3. Mr. Rowland Hussey Macy opened his *fancy dry goods* store in New York City on 6th Avenue and 14th street. His red star trademark became one of the most recognizable retail symbols.

   The number formed by the last two digits of the *Macy's* opening year is 10 more than six times the second digit of the year. A half of the number composed of the last two digits of the year is four times the second digit without 3. The first digit of the year is 1. What year did Mr. Rowland Hussey Macy open his store?

4. The department store *Dillard's* was founded in 1938 by William T. Dillard. Its corporate headquarter is in Little Rock, Arkansas. The chain has grown fast and opened stores almost in all US states.

   The largest number of *Dillard's* is in Texas followed by Florida. The *Dillard's* headquarter state, Arkansas, has less *Dillard's* than Texas and Florida. California has the lowest number of *Dillard's* compared to Texas, Florida, and Arkansas. How many *Dillard's* are in Texas, Florida, Arkansas, and California if the sums of *Dillard's* in any three states (Texas, Florida, Arkansas, and California) are equal to 107, 102, 68, 53?

5. The American chain of department stores *Sears* was founded by Richard Warren Sears and Alvah Curtis Roebuck as a mail ordering catalog company. Its first store was open in Indiana 33 years later. The headquarter was based at the Sears Tower in Chicago and is currently in Hoffman Estates, Illinois. *Sears* had the largest revenue of any retailer in the USA until *Walmart*, founded later than *Sears*, bit it.

   When was *Sears* founded? When was its first store opened? When was *Walmart* founded and got a larger revenue? The last row of the chart provides answers to all questions.

	19		19	19	19	
	77	58		___	___	1
1815		___	___	___	2009	

6. Sam Walton opened *Walton's 5 & 10* in 1950 in Arkansas that had been grown to the successful American discount department chain *Walmart* by 1962. The chain quickly expanded their location to other states by 1972 and to even more states by 1975.

   The number of US states that had *Walmart* by 1972 is the solution x to the equation $\frac{1}{x} = \frac{5}{2} - \frac{50 + \frac{3}{2}x}{20 + x}$, such that $|-x| = x$. The number of US states where *Walmart* was introduced between 1973 and 1975 is another solution to that equation for which $|x| = -x$. In both cases the absolute value of x is taken as the answer. Find the number of US states that had *Walmart* by 1972. Find the number of US states where *Walmart* was introduced between 1973 and 1975. Can you name these states?

7. The American discount department store chain *Stein Mart* was founded in 1908 by Sam Stein with a store in Greenville, Mississippi. Its headquarters was later moved to Jacksonville, Florida.

   Leadership plays the vital role in any business, and *Stein Mart* is a remarkable example. Indeed, it first stores carried general merchandise until Stein's son, Jake, took over the company and changed its directions to offering clothes, home décor, accessories, and shoes at discounted prices. Jake opened 120 new *Stein Mart* stores from 1977 to 1996, or, in other words, their number grew 41 times. How many *Stein Mart* stores were in 1977 and 1996?

8. ULTA Salon, Cosmetics & Fragrance, Inc. founded in 1990 is a chain of beauty superstores in the USA. Its headquarters is in Bolingbrook, Illinois.

   ULTA mails three types of coupons, $3.50 off any purchase of $10 or above, $5 off any purchase of $15 or above, or 20% off any purchase. Which option presented in the table below allows to save the most (in the corresponding price range)? When is it profitable to use 20% off any purchase coupon?

A.	$3.50 off any purchase of $10 or above, 20% off for any purchase above $15
B.	$5 off any purchase of $15 or above, 20% off for any purchase above $15
C.	$3.50 off any purchase of $10 or above, 20% off for any purchase above $25
D.	$5 off any purchase of $15 or above, 20% off for any purchase above $25
E.	20% off any purchase

9. A few years ago, John received a message from Verizon on Tuesday that indicated that a limited number of minutes was left. The new billing period would start in 10 days. Minutes on weekend were free

to use. John calculated that the average time for a phone conversation should be 12 minutes and 30 seconds until the beginning of the new period. Being careless, John talked for 21 minutes and 15 seconds on the first day. He understood that he could not reduce the duration of time he was talking on the phone immediately and decided to decrease the time gradually, the same amount of time each day. How much less did he talk each day if he talked even the very last day and had no minutes before the next billing period?

10. Each visa card has a four-digit PIN code. In how many ways digits from 0 to 9 can be arranged to form a PIN code if (a) no digits can be repeated; (b) digits can be repeated?

Answers

1. C
2. C
3. 1858
4. 57, 42, 8, 3
5. 1892, 1925, 1950, 1989
6. 5, 4
7. 3, 123
8. C, larger than $25
9. 2 min and 30 sec
10. 5040, 10,000

Solutions

1. 1st way: the problem can be easily solved without any mathematics if it is carefully read.

   2nd way: a mathematical solution. Let x be the original price of the dress, then $x = 0.5x + 58 \Rightarrow 0.5x = 58 \Rightarrow x = 116$.

2. Let us prepare a proportion: $\begin{matrix} \$100 & 80\% \\ x & 100\% \end{matrix} \Rightarrow x = \frac{100 \cdot 100}{80} = \$125$.

3. 1st way. Let a be the second digit of the year and b be the two-digit number formed with the last two digits, then

   $$\begin{cases} 6a + 10 = b \\ \frac{b}{2} = 4a - 3 \end{cases} \text{ or } \begin{cases} 6a + 10 = b \\ b = 8a - 6 \end{cases}. \text{ Thus, } 6a + 10 = 8a - 6 \text{ and } a = 8, b = 58.$$

   2nd way. The problem can be solved differently, noticing that the two-digit number formed with the last two digits is presented in two different ways from which immediately follows $6a + 10 = 8a - 6$.

Mr. Rowland Hussey Macy opened his "fancy dry goods" store in 1858.

4. Let t, f, a, and c be numbers of Dillard's stores Texas, Florida, Arkansas, and California. Then $\begin{cases} t + f + a = 107 \\ t + f + c = 102 \\ t + a + c = 68 \\ f + a + c = 53 \end{cases}$. Subtracting the second equation from the first one, we get $a - c = 5$, which becomes the fifth equation. Adding the third and fourth equations and subtraction the second equation, we get $2a + c = 19$, which in combination with the fifth equation, leads to $a = 8$ and $c = 3$. Substituting these values to the first and second equations, we obtain $f = 53 - 11 = 42$ and $t = 68 - 11 = 57$.

There were 57 Dillard's stores Texas, 42 in Florida, 8 in Arkansas, and 3 in California.

5. Looking carefully at the pattern, one can get,

	19		19		19		19	
	77		58		39		20		1
1815		1892		1950		1989		2009

and $1892 + 33 = 1925$.

6. The equation $\dfrac{1}{x} = \dfrac{5}{2} - \dfrac{50 + \frac{3}{2}x}{20 + x}$ or $\dfrac{1}{x} = \dfrac{5}{2} - \dfrac{100 + 3x}{2(20 + x)}$ can be simplified to $x^2 - x - 20 = 0$, that has two solutions $x = 5$ and $x = -4$. The first $x = 5$ satisfies $|-5| = 5$, and the second $x = -4$ satisfies $|-4| = -(-4) = 4$.

Arkansas, Kansas, Louisiana, Missouri, and Oklahoma had Walmart by 1972 and Walmart was introduced to Tennessee in 1973, Kentucky and Mississippi in 1974, and Texas in 1975.

7. Let x be the number of Stein Marts in 1977, then $41x$ is their number in 1996 and $41x - x = 120$. Then $x = 3$ stores in 1977 and 123 stores in 1996.

8. $3.50 off any $10 purchase offers 35% saving while $5 off any $15 purchase saves 33.3%; $20 < 33.3 < 35$.

Let us find when a 20% coupon is the best. If x is the purchase, then $0.2x > 5$ (20% coupon is better than $5 of saving) leads to $x > \$25$.

9. 10 days from Tuesday without weekends is 8 days. Using the arithmetic sequence formula (see Appendix I-SS), we obtain:

$$S_n = \frac{n}{2}(2a_1 + (n-1)d) \Rightarrow 8 \cdot 12.5 = \frac{8}{2}(2 \cdot 21.25 + 7d) \Rightarrow d = -2.5.$$

John talked 2 minutes and 30 seconds less each day.

10. If no digits can be repeated, then $10 \cdot 9 \cdot 8 \cdot 7 = 5040$; if digits are repeated, then $10 \cdot 10 \cdot 10 \cdot 10 = 10,000$ (see Appendix I-CR).

VII.4 Walt Disney Parks

The Walt Disney World is among the largest and most visited places in the world. It was open on October 1, 1971 in Lake Buena Vista, Florida, with the Magic Kingdom Park dedicated to fairy tales and Disney characters. EPCOT was open on October 1, 1982 to celebrate human achievement, technological innovations, and international culture. The third theme park, Disney's Hollywood Studios (Disney-MGM Studios until 2008), has welcomed visitors since May 1, 1989. Disney's Animal Kingdom with animal conservation theme has greeted guests since April 22, 1998.

The Walt Disney World Resort is also famous for its water parks, resort hotels, golf courses, shopping, dining, and entertainment centers.

1. The Walt Disney World Resort in Lake Buena Vista, Florida, covers x acres of land, where x is the solution to

$$7\log_x 2 + 4\log_x 5 + \log_x 47 = \log_x x + \log_x 125.$$

What is the area of the Walt Disney World Resort?

2. The areas of Magic Kingdom and Hollywood Studios are related to as 4 to 5, while the areas of Magic Kingdom and EPCOT are related to as 9 to 25. The area of EPCOT is 3/5 of the area of Animal Kingdom. The area of Animal Kingdom is 40 acres less than four times the area of Hollywood Studios. What is the area of each theme park?

3. How many theme parks does the Walt Disney World have if this number can be presented as $-\log_4 \log_2 \sqrt[2]{\sqrt[4]{\sqrt[8]{\sqrt[16]{16}}}}$? Solve without a calculator.

4. Spaceship Earth is the EPCOT iconic structure. It is the 18-story geodesic sphere that offers a 15-minute machine-themed ride back and forward in time to witness important achievements through

the history. Famous science fiction writer Ray Bradbury helped with its design.

Geometrically, Spaceship Earth is a dodecahedron with each of its 60 isosceles triangle faces divided into 16 smaller triangles, which, in turn, are sub-divided into four triangles, each of which is further divided into three isosceles triangles. What is the total number of all isosceles triangles?

5. Geometrically, Spaceship Earth of the EPCOT is a dodecahedron with each of its isosceles triangle faces divided into smaller triangles, which are sub-divided into four triangles, each of which is further divided into smaller isosceles triangle resulting in 11,520 isosceles triangles. Find the side of the smallest isosceles triangle if the diameter of the Spaceship Earth is 165 ft. Assume a spherical shape of Spaceship Earth.

6. Disney's Animal Kingdom theme park has the Tree of Life, a sculpted artificial baobab tree. If there were a cylinder with the height and diameter of the Tree of Life, then its surface area (with top and bottom) would be $8500\,\pi$ sq. ft. What are the height and diameter of The Tree of Life, if they are related to as 29 to 10?

7. Disney's Animal Kingdom is accredited by the World Association of Zoos and Aquariums. It is represented by the Tree of Life, a sculpted baobab tree. If there were a cylinder with the height and diameter of the Tree of Life, then its volume would be $90,625\,\pi$ cubic ft and its surface area would be $8500\,\pi$ sq. ft. What are the height and diameter of the Tree of Life? The cylinder is taller than wider and has the top and bottom.

8. How many resort hotels does the Walt Disney World Resort have if this number is equal to $81^{\frac{1}{4}-\frac{1}{2}\log_9 49}\cdot 1331^{\log_{121} 49}$? Solve without a calculator.

9. A golf can be played at $125^{\frac{1}{\log_4 25}}$ golf courses offered by the Walt Disney World Resort. Find the number of golf courses. Solve without a calculator.

10. How many water parks does the Walt Disney World Resort have if this number is $2\log_3 6 - \dfrac{1}{\log_2 \sqrt{3}}$? Solve without a calculator.

11. It takes 6 hours to fill in a water pool at the Disney's Typhoon Lagoon if two hoses are used. Due to a scheduled repair, the hoses were not used simultaneously. The first hose was used to fill the pool during the time needed for the second hose to fill one third of the pool if it worked alone. Then the first hose was turned off and the second hose was open during the time the first hose needed to fill the half of the pool if it worked alone. As a result, only 5/6 of the pool was filled in. How much time is needed for each hose to fill the pool if it works alone?

12. It takes 6 hours to fill in a water pool at the Disney's Typhoon Lagoon if two hoses are used. Due to a scheduled repair, the hoses were not used simultaneously. Only the first hose was used during the time the second hose needed to fill one third of the pool if it worked alone. Then only the second hose was open during the time needed for the first hose to fill the half of the pool if it worked alone. As a result, only the portion a of the pool was filled in. How much time does it take for each hose to fill the pool if it works alone? Find the values of a that lead to a unique solution, two solutions, or no solutions.

Answers

1. 30,080 acres
2. 108 acres, 300 acres, 135 acres, 500 acres
3. 4
4. 11,520
5. 4.14 ft
6. 145 ft, 50 ft
7. 145 ft, 50 ft
8. 21
9. 8
10. 2
11. 10 hours and 15 hours, 12 hours and 12 hours
12. No solutions when $0 < a < \sqrt{\frac{2}{3}}$, $x = \frac{3-\sqrt{6}}{6}$ when $a = \sqrt{\frac{2}{3}}$, two solutions when $\sqrt{\frac{2}{3}} < a < 1$

Solutions

1. The domain is $x > 0$, $x \neq 1$. Using the properties of logarithms $\log_a M + \log_a N = \log_a MN$ and $k \log_a M = \log_a M^k$ (see Appendix I-EL), the equation $7\log_x 2 + 4\log_x 5 + \log_x 47 = \log_x x + \log_x 125$ can be rewritten as

$$\log_x 2^7 + 4\log_x 5 + \log_x 47 =$$
$$\log_x x + 3\log_x 5 \Rightarrow \log_x 2^7 + \log_x 5 + \log_x 47 = 1 \Rightarrow \log_x 2^7 \cdot 5 \cdot 47 = 1.$$

Applying the relation between logarithmic and exponential functions (see Appendix I-EL), we get $x = 2^7 \cdot 5 \cdot 47 = 30080$.

The area of the World Disney World Resort is 30,080 acres.

2. Let x acres be the area of Animal Kingdom. Then the areas of EPCOT is $3x/5$, of Magic Kingdom is $9/25 \cdot (3x/5) = 27x/125$, and of Hollywood Studios is $5/4 \cdot (27x/125) = 27x/100$. Finally, from $4 \cdot$

$(27x/100) - x = 40$ follows $x = 500$.

The areas of the parks are: Magic Kingdom – 108 acres; EPCOT – 300 acres; Hollywood Studios – 135 acres; Animal Kingdom – 500 acres.

The areas of the parks are: Magic Kingdom – 108 acres; EPCOT – 300 acres; Hollywood Studios – 135 acres; Animal Kingdom – 500 acres.

3. Using the properties $\sqrt[k]{a} = a^{\frac{1}{k}}$, $(a^k)^m = a^{km}$, and $k\log_a M = \log_a M^k$ $k\log_a M = \log_a M^k$(see Appendix I-EL), we can simplify the expression as

$$-\log_4 \log_2 \sqrt[4]{\sqrt[8]{\sqrt[16]{16}}} = -\log_4 \log_2 16^{\frac{1}{1024}}$$

$$= -\log_4 \log_2 2^{\frac{1}{256}} = -\log_4 \frac{1}{256}$$

$$= -\log_4 4^{-4} = -(-4)\log_4 4 = 4$$

There are four theme parks in the World Disney World.

4. $60 \cdot 16 \cdot 4 \cdot 3 = 11{,}520$ isosceles triangles.

   Remark. There are only 11,324 silvered facets. Some of those triangles are missing partially or fully due to supports and doors. All triangles are slightly curly to make the construction look like a sphere.

5. Using the formula for the surface area of a sphere $S = 4\pi r^2$ (see Appendix I-VS), we obtain the surface area of the Spaceship Earth as $S = 4\pi (165/2)^2 \approx 85{,}529.86$ sq. ft. The area A of the triangle is the surface area of the Spaceship Earth divided by the number of triangles, which is $A = 7.42$ sq. ft. Then from $A = \frac{a^2\sqrt{3}}{4} \Rightarrow a \cong 4.14$ft.

   The size of the smallest triangle is approximately 4.14 ft.

6. The radius r and the height h of the tree are related to as $\frac{h}{r} = \frac{29}{5} \Rightarrow$ $h = \frac{29}{5}r$. Its surface area S (see Appendix I-VS) is $S = 2\pi r^2 + 2\pi rh = 8500\pi \Rightarrow 2\pi r^2 + 2\pi r\frac{29}{5}r = 8500\pi \Rightarrow 2\pi\frac{34}{5}r^2 = 8500\pi \Rightarrow r = 25$.

   The Tree of Life is 145-ft tall with a 50-ft diameter.

7. The volume of a cylinder (see Appendix I-VS) is $V = \pi r^2 h = 90{,}625\pi$, from which its height is found as $h = \frac{90{,}625}{r^2}$. From the formula of the surface area S of a cylinder $S = 2\pi r^2 + 2\pi rh = 8500\pi$, we get $2\pi r^2 +$

$2\pi r \frac{90,625}{r^2} = 8500\pi$ that leads to a cubic equation $r^3 - 4250r + 90,625 = 0$.

The cubic equation can be solved in different ways. One of them is considering all factors of 90,625 (because the leading coefficient is 1). The prime factorization of $90,625 = 5^5 \cdot 29$. Checking 5 and then 25, we can see that 25 satisfies the cubic equation and reduce it to $(r - 25)(r^2 + 25r - 3625) = 0$. Solving the quadratic equation $r^2 + 25r - 3625 = 0$, we get 25, $\frac{-25 - 55\sqrt{5}}{2}$, and $\frac{-25 + 55\sqrt{5}}{2}$ as three roots of the cubic equation $r^3 - 4250r + 90,625 = 0$. The second root $\frac{-25 - 55\sqrt{5}}{2}$ does not work because it is negative, $\frac{-25 + 55\sqrt{5}}{2}$ does not work because then the height will be approximately 37.75 ft, which is smaller than the diameter. Thus, only $r = 25$ satisfies the problem.

The Tree of Life is 145-ft-tall with a 50-ft-diameter.

8. Using properties of exponential and logarithmic functions (see Appendix I-EL) $k \log_a M = \log_a M^k$ and $a^{\log_a M} = M$, the expression can be simplified to $81^{\frac{1}{4} - \frac{1}{2}\log_9 49} \cdot 1331^{\log_{121} 49} = \frac{81^{\frac{1}{4}}}{81^{\frac{1}{2}\log_9 49}} \cdot 11^{3 \log_{11} 7} = \frac{3}{49} \cdot 7^3 = 21$

   The World Disney World resort has 21 hotels.

9. Using the properties of exponential and logarithmic functions (see Appendix I-EL) $k \log_a M = \log_a M^k$, $\log_a M = \frac{1}{\log_M a}$, and $a^{\log_a M} = M$, the expression can be simplified to $125^{\frac{1}{\log_4 25}} = 5^{3\log_5 2} = 2^3 = 8$.

   The World Disney World resort has 8 golf courses.

10. Since $\log_a M - \log_a N = \log_a \frac{M}{N}$, $k \log_a M = \log_a M^k$, and $\log_a M = \frac{1}{\log_M a}$ (see Appendix I-EL), the expression can be simplified as $2\log_3 6 - \frac{1}{\log_2 \sqrt{3}} = \log_3 36 - \frac{1}{\log_4 3} = \log_3 36 - \log_3 4 = \log_3 9 = 2$.

   The World Disney World resort has 2 water parks.

11. Let x and y be the productivity of each hose per hour and 1 be the entire job to fill in the pool. Then $1/x$ and $1/y$ is the number of hours needed for each hose to fill the pool if it works alone. The number of the hours during which the second hose fills one third of the pool if it works alone is $1/3 \cdot 1/y$, and $1/3 \cdot 1/y \cdot x$ is the portion of the pool filled by the first hose during this time. Hence,

$$\begin{cases} 6(x+y)=1 \\ \frac{1}{3}\frac{1}{y}x + \frac{1}{2}\frac{1}{x}y = \frac{5}{6} \end{cases} \Rightarrow \begin{cases} y = \frac{1}{6} - x \\ \dfrac{x}{\frac{1}{2}-3x} + \dfrac{\frac{1}{6}-x}{2x} = \frac{5}{6} \end{cases} \Rightarrow \begin{cases} y = \frac{1}{6} - x \\ 120x^2 - 22x + 1 = 0 \end{cases}$$

$$\Rightarrow \begin{cases} x = \frac{1}{10}, \ x = \frac{1}{12} \\ y = \frac{1}{15}, \ y = \frac{1}{12} \end{cases}.$$

The problem has two solutions: 10 and 15 hours or 12 and 12 hours.

12. It is obvious that $0 < a \le 1$. Let x and y be the productivity of each hose in 1 hour, then $1/x$ and $1/y$ are the number of hours each hose takes to fill the pool if it works alone and

$$\begin{cases} 6(x+y)=1 \\ \frac{1}{3}\frac{1}{y}x + \frac{1}{2}\frac{1}{x}y = a \end{cases} \Rightarrow \begin{cases} y = \frac{1}{6} - x \\ \dfrac{x}{\frac{1}{2}-3x} + \dfrac{\frac{1}{6}-x}{2x} = a \end{cases} \Rightarrow \begin{cases} y = \frac{1}{6} - x \\ x^2\left(a + \frac{5}{6}\right) - x\left(\frac{a}{6} + \frac{1}{6}\right) + \frac{1}{72} = 0 \end{cases}$$

$$\Rightarrow \begin{cases} y = \frac{1}{6} - x \\ 12x^2(6a+5) - 12x(a+1) + 1 = 0 \end{cases}$$

$$x = \frac{6(a+1) \pm \sqrt{36(a+1)^2 - 12(6a+5)}}{12(6a+5)}.$$

A unique solution (see Appendix I-QE) occurs if $D = 36(a+1)^2 - 12(6a+5) = 0 \Rightarrow 3a^2 = 2 \Rightarrow a = \pm\sqrt{\frac{2}{3}}$, the solution is $x = \frac{3-\sqrt{6}}{6}$.

There are no solutions if $-\sqrt{\frac{2}{3}} < a < \sqrt{\frac{2}{3}}$, which in combination with $0 < a < 1$ leads $0 < a < \sqrt{\frac{2}{3}}$.

There are two solutions if $\sqrt{\frac{2}{3}} < a < 1$.

Appendix I

Appendix I presents a set of algebraic formulas and properties used in this book. It is not intended to be considered as a systematic mathematical reference. The topics needed for a review can be easily found using a subsection title. A reference to sections of Appendix I are provided in the *problem solution* section. For easy search, they contain the first letters of the required formula, that is, Appendix I-QE reviews quadratic equations.

AF. Algebraic Formulas and Identities

Basic identities:

$$(a + b)^2 = a^2 + 2ab + b^2 \qquad (a - b)^2 = a^2 - 2ab + b^2 \qquad a^2 - b^2 = (a - b)(a + b)$$

$$(a + b)^3 = a^3 + 3a^2b + 3ab^2 + b^3 \quad (a - b)^3 = a^3 - 3a^2b + 3ab^2 - b^3$$

$$a^3 + b^3 = (a + b)(a^2 - ab + b^2) \quad a^3 - b^3 = (a - b)(a^2 + ab + b^2)$$

The *binomial theorem* or the *binomial expansion* is the algebraic presentation of the power of a binomial $(a + b)$ as a polynomial:

$$(a + b)^n = \sum_{k=0}^{n} C_n^k a^{n-k}b^k = C_n^0 a^n + C_n^1 a^{n-1}b^1 + \ldots + C_n^k a^{n-k}b^k + \ldots + C_n^{n-1}ab^{n-1}$$

$$+ \ldots + C_n^n b^n,$$

where $C_n^k = \dfrac{n!}{k!\,(n-k)!}$, $n! = \begin{cases} 1 \cdot 2 \cdot 3 \cdot \ldots \cdot n, & n \geq 1 \\ 1 & n = 0 \end{cases}$.

Reduction of $\sqrt{a \pm \sqrt{b}}$ *to* $\dfrac{\sqrt{m} \pm \sqrt{n}}{\sqrt{2}}$ is possible if there are two positive numbers m and n with the product b and the sum $2a$. Steps:

1. Multiple and divide $a \pm \sqrt{b}$ by 2: $\sqrt{a \pm \sqrt{b}} = \sqrt{\dfrac{2(a \pm \sqrt{b})}{2}} = \sqrt{\dfrac{2a \pm 2\sqrt{b}}{2}}$.

2. Find m and n such as $mn = b$, $m + n = 2a$:
$$\sqrt{a \pm \sqrt{b}} = \sqrt{\frac{2a \pm 2\sqrt{b}}{2}} = \sqrt{\frac{m + n \pm 2\sqrt{mn}}{2}}.$$

3. Use exponential properties, $m = \sqrt{m^2} = (\sqrt{m})^2$ and $\sqrt{nm} = \sqrt{n}\sqrt{m}$:
$$\sqrt{a \pm \sqrt{b}} = \sqrt{\frac{m + n \pm 2\sqrt{mn}}{2}} = \sqrt{\frac{(\sqrt{m})^2 \pm 2\sqrt{m}\sqrt{n} + (\sqrt{n})^2}{2}}.$$

4. Use the square of a sum or difference formula $(a \pm b)^2 = a^2 \pm 2ab + b^2$:
$$\sqrt{a \pm \sqrt{b}} = \sqrt{\frac{(\sqrt{m})^2 \pm 2\sqrt{m}\sqrt{n} + (\sqrt{n})^2}{2}} = \sqrt{\frac{(\sqrt{m} \pm \sqrt{n})^2}{2}}.$$

5. Simplify: $\sqrt{a + \sqrt{b}} = \sqrt{\frac{(\sqrt{m} + \sqrt{n})^2}{2}} = \frac{\sqrt{m} + \sqrt{n}}{\sqrt{2}}$,

$$\sqrt{a - \sqrt{b}} = \sqrt{\frac{(\sqrt{m} - \sqrt{n})^2}{2}} = \begin{cases} \dfrac{\sqrt{m} - \sqrt{n}}{\sqrt{2}}, & m \geq n \\[2mm] \dfrac{\sqrt{n} - \sqrt{m}}{\sqrt{2}}, & m < n \end{cases}.$$

Example: Simplify $\sqrt{5 - \sqrt{21}}$.

Since $3 \cdot 7 = 21$ and $3 + 7 = 10 = 2 \cdot 7$, we obtain

$$\sqrt{5 - \sqrt{21}} = \sqrt{\frac{10 - 2\sqrt{21}}{2}} = \sqrt{\frac{(\sqrt{7})^2 - 2\sqrt{7 \cdot 3} + (\sqrt{3})^2}{2}} = \sqrt{\frac{(\sqrt{7} - \sqrt{3})^2}{2}}$$
$$= \frac{\sqrt{7} - \sqrt{3}}{\sqrt{2}}.$$

CP. Circle

r – the *inradius*, the radius of the incircle, the largest circle that will fit inside a polygon.

R – the *circumradius*, the radius of the circumscribed circle of a polygon.

$C = 2\pi r$ – the *circumference* of a circle.

$A = \pi r^2$ – the *area* of a circle.

A *tangent* to a circle – a straight line outside a circle which touches the circle at only one point.

The Tangent to a Circle Theorem: A line is tangent to a circle if and only if it is perpendicular to the radius drawn to the point of tangency.

The Two Tangents Theorem: Two tangent segments to a circle from the same external point have the same length.

Inscribed Quadrilateral Theorem: A quadrilateral can be inscribed in a circle if and only if its opposite angles are supplementary.

Circumscribed Quadrilateral Theorem: A circle can be inscribed in a quadrilateral if and only the sums of its opposite sides are the same.

AV. Absolute Value

Basic formulas:

$$|x| = \begin{cases} x, & x > 0 \\ 0, & x = 0. \\ -x, & x < 0 \end{cases}$$

$$|x| \geq a \Rightarrow x \leq -a \text{ and } x \geq a, \ a \geq 0.$$

$$|x| \leq a \Rightarrow -a \leq x \leq a, \ a \geq 0.$$

CR. Counting Rules

Multiplication rule – the number of different ways of selecting one element from each of k sets of elements with the sizes $n_1, n_2, n_3,..., n_k$ of the sets is $n_1 \cdot n_2 \cdot n_3 \cdot ... \cdot n_k$.

Permutation rule – the number of selecting k elements from a set of n elements if the selection order matters is $P_n^k = \frac{n!}{(n-k)!}, n! = \begin{cases} 1 \cdot 2 \cdot 3 \cdot ... \cdot n, & n \geq 1 \\ 1, & n = 0 \end{cases}$.

Combination rule – the number of selecting k elements from a set of n elements without considering the order of selection is $C(n, k) = C_n^k = \frac{n!}{k!(n-k)}$.

EL. Exponential and Logarithmic Functions

An *Exponential Function* $y = a^x$, $a > 0$, $a \neq 1$. Domain: $x \in (-\infty, \infty)$, Range: $y \in (0, \infty)$.

A *Logarithmic Function* $y = \log_a x$, $a > 0$, $a \neq 1$. Domain: $x \in (0, \infty)$, Range: $y \in (-\infty, \infty)$.

Relations between logarithmic and exponential functions: $b^a = e^{a \ln b}$, $\log_b b^a = a$, $\log_a b = c \Leftrightarrow a^c = b$.

Properties of exponential functions:

$$a^0 = 1 \qquad a^1 = a \qquad a^{-n} = 1/a^n \quad \sqrt[n]{a} = a^{\frac{1}{n}} \qquad (a^n)^m = a^{nm}$$
$$a^n a^m = a^{n+m} \quad a^n/a^m = a^{n-m} \quad (ab)^n = a^n b^n \quad (a/b)^n = a^n/b^n$$

Properties of logarithmic functions:

$$\log_a MN = \log_a M + \log_a N \quad \log_a \frac{M}{N} = \log_a M - \log_a N$$
$$\log_{a^k} M^r = \frac{r}{k}\log_a M \qquad\qquad \log_a M = \frac{\log_b M}{\log_b a} \qquad\qquad \log_a M = \frac{1}{\log_M a}$$

F. Functions

A *function* $y = f(x)$ – a relation between two sets, such that each element x of the first set has exactly one element y of the second set.

The *Domain* of a function – the set of all values of x (the first set).

The *Range* of a function – the set of all values of y (the second set).

A function is *odd* if $f(x) = -f(-x)$. An odd function is symmetric about the origin.

A function is *even* if $f(x) = f(-x)$. An even function is symmetric about the y-axis.

MU. Measurement Units

Units of length

1 mi = 1.609344 km	1 km = 0.621371 mi = 1093.61 yard	
1 ft = 0.3048 m	1 m = 3.28084 ft	1 ft = 12 in
1 in = 2.54 cm	1 cm = 0.393701 in	
1 nautical mile = 1.852 km	1 nautical mile = 1.15 mi	

Units of weight

1 kg = 2.20462 lb   1 lb = 0.45359 kg   1 oz = 0.0625 lb

QC. Quadratic Curves

$Ax^2 + By^2 + Cxy + Dx + Ey + D = 0$, $A + B + C \neq 0$ – a general equation of quadratic curves.
  Special cases:

1. $\frac{(x-a)^2}{h^2} + \frac{(y-b)^2}{k^2} = 1$ – an ellipse with the center at (a, b), passes through $(a - h, b)$, $(a + h, b)$, $(a, b + k)$, $(a, b - k)$.
   $(x - a)^2 + (y - b)^2 = r^2$ – a circle with the center at (a, b) and radius r.

2. $\frac{(x-a)^2}{h^2} - \frac{(y-b)^2}{k^2} = 1$ – a hyperbola opens left and right.
   $\frac{(y-b)^2}{k^2} - \frac{(x-a)^2}{h^2} = 1$ – a hyperbola opens up and down.

3. $y = a(x - b)^2 + c$, $a \neq 0$ – a parabola with the vertex at (b, c), symmetric about $x = b$, and $\begin{cases} \text{opens up,} & a > 0 \\ \text{opens down,} & a < 0 \end{cases}$.

QE. Quadratic Equation

$ax^2 + bx + c = 0$, $a \neq 0$ – a *quadratic equation*.

$D = b^2 - 4ac$ – the discriminant of a quadratic equation $ax^2 + bx + c = 0$, $a \neq 0$.

The *discriminant* describes solutions of a quadratic equation:
$$\begin{cases} \text{two distinct real solutions,} & D > 0 \\ \text{two repeated real solutions,} & D = 0. \\ \quad\text{no real solutions,} & D < 0 \end{cases}$$

$x = \frac{-b \pm \sqrt{b^2 - 4ac}}{2a}$ – solutions of the quadratic equation $ax^2 + bx + c = 0$, $a \neq 0$.

$x = -\frac{p}{2} \pm \sqrt{\left(\frac{p}{2}\right)^2 - q}$ – solutions of the quadratic equation $x^2 + px + q = 0$, $D = \left(\frac{p}{2}\right)^2 - q$.

The *Viète theorem*: Two roots x_1 and x_2 of the quadratic equation $x^2 + px + q = 0$ satisfy both $x_1 + x_2 = -p$ and $x_1 \cdot x_2 = q$.

PE. Polynomial Equations with Real Coefficients

$p_n(x) = a_n x^n + a_{n-1} x^{n-1} + \ldots + a_2 x^2 + a_1 x + a_0 = 0$, $a_n \neq 0$ – a polynomial equation of the n-th degree with real coefficients $a_n, a_{n-1}, \ldots, a_1, a_0$.

A polynomial equation $p_n(x) = 0$ of the n-th degree has n zeroes (solutions). The product of all zeroes of $p_n(x) = 0$ is a_0/a_n.

The *number of real solutions* of the polynomial equation is equal to n or any nonnegative number less than n by an even number (obtained by subtracting an even number from n, e.g., $n - 2$, $n - 4$, etc.).

Descartes rule of sign – the number of *positive* real zeroes in $p_n(x) = 0$ is equal to or any nonnegative number less than (by an even number) the number of changes in the sign of the coefficients of $p_n(x)$. The number of its *negative* real zeroes is equal to or less than (by an even number) the number of changes in the sign of the coefficients of $p_n(-x)$.

PF. Prime Factorization

A *prime* number – a number that has exactly two factors, 1 and the number itself, e.g., 2, 5.

A *composite* number – a number that has more than two factors.

Co-prime numbers – two numbers that do not have any common factors other than 1, e.g., 5 and 16, 18 and 49.

Prime factorization of a number – representation of the number as a product of only prime numbers.

The Fundamental Theorem of Arithmetic – any positive integer $a > 1$ is either a prime number or can be uniquely presented as the product of prime numbers disregarding the order of the factors.

The *number of divisors* of a positive integer $a > 1$ is equal to the product $(\alpha_1 + 1)(\alpha_2 + 1) \cdot \ldots \cdot (\alpha_n + 1)$, where $\alpha_1, \alpha_2, \ldots, \alpha_n$ are exponents of the unique prime factorization $a = p_1^{\alpha_1} \cdot p_2^{\alpha_2} \cdot \ldots \cdot p_n^{\alpha_n}$ and $p_1, p_2, \ldots, p_n$ are distinct prime factors of a.

PT. Pythagorean Triples

(a, b, c) – Pythagorean triple if $c^2 = a^2 + b^2$.
Some formulas for a Pythagorean triple (a, b, c):
$(2n^2, n^2 - 1, n^2 + 1)$ for any n,

$\left(n, \dfrac{n^2 - 1}{2}, \dfrac{n^2 + 1}{2}\right)$ if n is even and $\left(n, \dfrac{n^2}{4} - 1, \dfrac{n^2}{4} + 1\right)$ if n is odd.

SS. Sequences and Series

a_1 – the first term of an arithmetic sequence.
a_n – the n-th term of an arithmetic sequence.
d – the constant difference of an arithmetic sequence.
b_1 – the first term of a geometric sequence.
b_n – the n-th term of a geometric sequence.
q – the constant ratio of a geometric sequence.
S_n – the sum of the first n terms of a sequence.
S – the sum of infinitely many terms.
An *arithmetic sequence* $\{a_n\}$ – $a_n = a_{n-1} + d$, $a_n = a_1 + d(n - 1)$, $S_n = \frac{a_1 + a_n}{2}n$,
$S_n = \frac{2a_1 + d(n - 1)}{2}n$, $n = 1, 2, 3, 4, \ldots$.

A *geometric sequence* $\{b_n\}$ – $b_n = b_{n-1}q$ or $b_n = b_1 q^{n-1}$, $S_n = b_1 \frac{1 - q^n}{1 - q}$, $n = 1, 2, 3,$

$4, \ldots, S = b_1 \frac{1}{1 - q}$ if $|q| < 1$.
The *Fibonacci sequence* $\{0, 1, 1, 2, 3, 5, 8, \ldots\}$ – $F_{n + 2} = F_{n + 1} + F_n$, $n = 2, 3,$
$4, \ldots, F_0 = 0, F_1 = 1$.
The *Lucas sequence* $\{2, 1, 3, 4, 7, 11, \ldots\}$ – $L_{n+2} = L_{n+1} + L_n$, $n = 2, 3, 4, \ldots, L_0 = 2,$
$L_1 = 1$.

Triangular numbers – the number of dots needed to draw a triangle,
$T_n = \frac{n(n + 1)}{2}$.

T. Triangles

a, b, c – three sides of a triangle opposite to $\angle A$, $\angle B$, and $\angle C$, respectively.
P – the perimeter of a triangle.
p – the half perimeter of a triangle.
A – the area of a triangle.
r – the inradius, the radius of the inscribed circle.
R – the circumradius, the radius of the circumscribed circle.
l_a – the angle bisector to the opposite side a – a line segment from a vertex to the opposite side that divides the angle into two congruent angles.
m_a – the median to the side a – the line segment from the vertex A to the midpoint of the opposite side a.
h_a – the height to the side a – the line segment from the vertex A perpendicular to the opposite side a.
a_b (connected to the side b) and a_c (connected to the side c) – the lengths of the two segments that the side a is divided into by the angle bisector, the median, or the height to the side a.

Properties:
The sum of the lengths of two sides of a triangle is greater than the length of the third side.
The point of intersection of three angle bisectors is its incenter.
The incenter is equidistant from the sides of a triangle.
The median, bisector, and height from the vertex in an isosceles triangle coincide.
All medians, bisectors, and heights in an equilateral triangle coincide.

Types of Triangles by sides:
A *scalene triangle* – all sides have different lengths.
An *isosceles triangle* – two sides are of equal length.
An *equilateral triangle* – the three sides have the same length.

Types of Triangles by internal angles:
An *acute triangle* – all interior angles are less than 90°. $a^2 + b^2 > c^2$, c is the length of the longest side.
A *right triangle* – one of its interior angles is a right angle or 90°. $a^2 + b^2 = c^2$, c is the length of the longest side.
An *obtuse triangle* – one of its interior angles is greater than 90, c is the length of the longest side.

Relations between sides and angles:
The *law of cosine:* $c^2 = a^2 + b^2 - 2ab \cos\angle C$.
The *law of sines:* $\frac{\sin \angle A}{a} = \frac{\sin \angle B}{b} = \frac{\sin \angle C}{c} = 2R$.

Area of a triangle:
$$A = \frac{bh_b}{2}, \quad A = \sqrt{p(p-a)(p-b)(p-c)}, \quad A = pr = \frac{a+b+c}{2}r, \quad A = \frac{abc}{4R},$$
$$A = \sqrt{p \cdot (p-a) \cdot (p-b) \cdot (p-c)} = \sqrt{\frac{a+b+c}{2} \cdot \frac{-a+b+c}{2} \cdot \frac{a-b+c}{2} \cdot \frac{a+b-c}{2}} \quad - \text{ the Heron's}$$
Formula.

Bisector, Median, Height. Useful formulas:
The length of a *median:* $4m_a^2 = 2b^2 + 2c^2 - a^2$ – the Apollonius' Formula.
Relations between sides and medians: $\frac{9}{4}a^2 = 2(m_b^2 + m_c^2) - m_a^2$, $3(a^2 + b^2 + c^2) = 4(m_a^2 + m_b^2 + m_c^2)$.
The *angle bisector theorem:* $\frac{a_b}{a_c} = \frac{c}{b}$.
The length of an *angle bisector:* $l_a^2 = bc - \frac{a^2bc}{(b+c)^2}$, $l_a^2 = \frac{4bcp(p-a)}{(b+c)^2}$, $l_a^2 = cb - a_c a_b$.
The relation between the *inradius* and three *heights:* $\frac{1}{h_a} + \frac{1}{h_b} + \frac{1}{h_c} = \frac{1}{r}$.

TF. Trigonometry

Basic Trigonometric Identities:

$$\sin^2 A + \cos^2 A = 1$$
$$1 + \tan^2 A = \sec^2 A$$
$$1 + \cot^2 A = \csc^2 A$$
$$\sin 2A = 2\sin A \cos A$$
$$\cos 2A = \cos^2 A - \sin^2 A$$

$$\sin(A - B) = \sin A \cos B - \cos A \sin B$$
$$\sin(A + B) = \sin A \cos B + \cos A \sin B$$
$$\cos(A - B) = \cos A \cos B + \sin A \sin B$$
$$\cos(A + B) = \cos A \cos B - \sin A \sin B$$

VS. Volume, Surface Area

r – the radius.
h – the height.
S – the surface area.
V – the volume.
A sphere: $V = \frac{4}{3}\pi r^3$, $S = 4\pi r^2$.
A cylinder: $V = \pi r^2 h$, $S = 2\pi r^2 + 2\pi rh$.
A cone: $V = \frac{1}{3}\pi r^2 h$, $S = 2\pi r^2 + 2\pi r\sqrt{h^2 + r^2}$.

Appendix II

The below table lists basic mathematical topics with links to matching problems. Only the highest level of the mathematical topic for a referred problem is indicated. For example, solving a quadratic equation can be a main content of one problem and it is just auxiliary for another. Then only the first problem will be referred under the *equation quadratic*. Each problem is cited with three numbers separated by periods to represent the numbers of a chapter, its section, and a problem in the section. For instance, V.3.1 is the first problem from the third section of the fifth chapter.

Mathematical Topic	Related Problems
Absolute Value	II.1.2, III.3.1, VI.1.6
Algebraic Identities	I.2.1, I.4.1, I.4.3, II.2.4, II.2.5, II.4.4–.5, III.1.1–.2, III.1.7, III.2.2, V.1.4, V.3.2, V.4.4, VI.1.11
Area	I.1.2, II.1.4, II.1.10, II.4.1, III.1.8–.9, III.2.2
Average, Median	I.1.16, II.1.7, II.2.10, II.2.12, IV.3.3, V.2.10
Circle	I.1.8, I.3.3, I.3.4, I.4.4, II.1.4, II.2.14, II.3.1, II.3.2, II.3.11, II.4.9, II.4.10, III.1.16, III.1.18, III.2.2, IV.2.10–.11, V.1.10–.11, V.1.15, VI.1.1
Curve	II.3.8, II.4.1, V.1.6, VI.3.10
Equation Linear	I.1.15, II.1.1, II.2.10, II.3.9, II.4.4, II.4.8, III.1.8–.9, III.3.9, III.3.15, IV.1.1, IV.1.4, IV.2.1, IV.2.6–.7, IV.3.2, IV.3.12, IV.3.16–.17, V.2.9, V.3.3–.4, V.4.8–.9, V.4.12, VI.4.5, VII.1.1–.10, VII.2.1–.6, VII.3.1–.2, VII.3.7–.9, VII.4.2
Equation Nonlinear	I.1.6, I.1.7, I.2.5, I.2.8, I.2.11, I.3.9, I.4.6, II.1.6, III.1.1–.5, III.2.6–.7, III.2.10, III.3.10–.12, IV.1.9, IV.2.15, IV.3.1–.2, IV.3.6, IV.3.15, V.3.10, V.4.6, V.4.9, VI.4.1–.2, VI.4.11, VII.3.6
Equation Quadratic	I.2.10, I.2.12, 2.15, I.3.11, I.4.5, II.1.3, V.1.1–.2, V.1.12–.13, V.4.3
Functions, Domain, Range, Operations	II.1.2, II.1.5, II.3.5, V.2.6–.8, V.3.5, V.4.6, V.4.11, VI.1.6–.7, VI.3.10, VII.2.10
Function Logarithmic, Exponential	I.2.2, I.3.6, I.3.7, I.4.7, II.1.5, III.2.7, III.3.2, V.3.7, V.4.10, VI.1.3, VI.1.7, VI.4.5, VII.2.7, VII.2.9, VII.4.1, VII.4.3, VII.2.11, VII.4.1, VII.4.3, VII.4.8–.9
Geometric Figures 3-D	II.2.14, II.3.4, V.1.9, V.1.11, V.2.11–.14, VII.4.4–.7
Inequalities	I.1.6, I.1.14, I.3.10, III.1.6
Prime Factorization Divisors, Multiples	I.1.4, I.1.16, I.2,7, I.4.2, III.2.4, III.3.2, IV.3.14, V.1.1, V.1.8, V.2.1, V.4.7–.8, VI.1.12, VI.2.5, VI.3.6
Pythagorean Theorem	I.2.4, I.4.8, II.1.4, V.3.8, V.4.2

(Continued)

Mathematical Topic	Related Problems
Sequence Arithmetic	I.3.8, I.4.4, I.4.9, II.1.9, II.3.6, III.2.8, III.3.5, III.3.7, IV.1.9–.10, V.1.7, V.1.14, V.2.2, V.2.5, V.3.8–.9, V.4.5, VI.3.5, VI.3.9, VII.3.9
Sequence Infinite	I.4.2, IV.3.5, VI.2.1, VI.4.3
Sequence Geometric	I.2.13, II.1.8, II.1.11, II.3.6, III.3.6, V.1.7, V.2.4
System of Linear Equations	I.1.9, I.1.10, I.1.12, I.1.13, I.3.5, II.1.12, II.2.6, II.2.7, II.2.8, II.2.9, II.2.12, II.3.8, II.3.10, III.1.10–.13, III.1.15, III.2.5, III.2.9, III.3.11, III.3.14, IV.1.2–.4, IV.1.6, IV.1.10–.11, IV.3.8, IV.3.11, V.1.3, V.2.10, V.3.1, VI.1.2, VI.2.3, VI.4.7–.10, VII.3.3–.4, VII.4.2
System of Nonlinear Equations	I.1.1, I.1.5, I.2.3, I.2.6, I.2.9, I.4.1, I.4.10, II.1.10, II.2.1, II.2.2, II.2.3, II.4.3, IV.2.8–.9, IV.3.10, V.1.10, V.4.1, VI.3.4, VII.4.11–.12
Triangle, Properties	I.1.2, I.2.4, I.3.3, I.3.4, I.4.4, I.4.8, II.2.11, II.3.1, II.3.11, II.4.1–.7, II.4.9, IV.2.2–.5, IV.2.8–.11, IV.2.13, V.1.10

Appendix III

This Appendix matches all US states with corresponding problems where the state is mentioned. Link to each problem contains three numbers separated by a period that represent the numbers of a chapter, its section, and a problem in the section. For instance, V.3.1 is the first problem from the third section of the fifth chapter.

Some states are cited in many problems, others are mentioned just in a few problems. Each US state has rich heritage that cannot be covered in one book. The authors do not attempt to give a comprehensive story of a state but rather put their emphasis on mathematical problems. They try to maintain a delicate balance between a brief story and related problems. The presented collection of informative problems reflects the authors' most frequent travel and is carefully designed to make the book attractive, educational, and not too large.

US State	Related Problems
Alabama (AL)	I.1.11, I.4.7, II.1.2, IV.3.8
Alaska (AK)	I.1.2, II.1.11, II.2.1, II.2.2, II.2.12, II.3.9, II.4.4–.5, III.1.8, III.1.9, III.2.5, VII.2.7
Arizona (AZ)	III.1.8, III.1.9, III.1.18, III.1.19, III.3.3–.5, IV.1.1–.4, IV.1.7, IV.1.9, IV.3.1, IV.3.8, V.1.12, V.1.15, V.4.3
Arkansas (AR)	I.1.11, II.1.2, IV.3.8, VII.2.6, VII.3.4, VII.3.6
California (CA)	II.2.2, II.2.9–.13, II.4.10, III.1.2, III.1.12, III.2.2, III.3.1, III.3.5, III.3.7, III.9–.10, IV.3.6, IV.3.8–.9, V.1.14, V.2.8, V.3.2, V.3.3, V.4.1, V.4.6, V.4.7, V.4.9, VII.2.1–.4, VII.3.4
Colorado (CO)	I.1.11, I.3.11, II.1.3, II.1.6, II.1.10, II.2.4, II.2.5, II.2.9, II.3.9, III.1.7, III.1.13–.14, III.1.17–.19, III.3.1, IV.1.6, IV.2.1, IV.2.13, IV.3.2–.4, V.1.4, V.4.2, V.4.5
Connecticut (CT)	I.4.6, I.4.10, III.3.1, VI.2.1
Delaware (DE)	I.1.3, I.4.6, II.1.5, III.1.6
Florida (FL)	I.4.4, I.4.6, I.4.9, II.1.4, II.3.1, II.4.1, III.3.6–.6, IV.3.6, IV.3.8, IV.3.9, V.1.6–.8, V.4.3, VII.2.1, VII.3.4, VII.3.7, VII.4.1–.12
Georgia (GA)	I.3.11, I.4.6, II.1.2, II.1.8, II.1.9, III.3.1, V.1.12, V.4.6, VI.3.6–.7, VII.2.4
Hawaii (HI)	I.1.2, I.1.3, I.3.11, II.2.6, II.3.5–.6, V.4.3, VII.2.7
Idaho (ID)	I.4.7, II.3.2, III.1.10–.12, III.3.1, III.4–.5, V.4.3
Illinois (IL)	I.1.4, I.1.10, I.4.6, II.1.2, II.1.7, IV.3.8, IV.3.14, V.1.3–.5, V.4.11, VII.2.5, VII.3.5, VII.3.8

(Continued)

US State	Related Problems
Indiana (IN)	VII.2.5
Iowa (IA)	I.4.3, I.4.6, II.1.2, V.3.2–.3
Kansas (KS)	II.1.2, II.2.14, IV.2.14, V.2.10, VII.3.6
Kentucky (KY)	II.1.2, III.3.5, IV.1.11, VII.3.6
Louisiana (LA)	I.4.6, I.1.9, I.1.10, II.1.11, II.2.13, III.3.11, IV.3.8, V.2.9, VII.3.6
Maine (ME)	I.4.4, I.4.7, III.3.1, III.3.5–.6, IV.1.8, IV.2.1, IV.2.12, IV.3.9, V.1.14
Maryland (MD)	I.3.8, I.4.6, III.3.1, III.3.7, VI.2.2
Massachusetts (MA)	I.4.3, I.4.4, I.4.10, II.4.6–.9, III.3.1, IV.3.12, IV.3.16–.17, V.2.2, VI.1.11–.12, VI.2.2, VI.3.3
Michigan (MI)	I.4.6, I.4.7, II.1.7, II.1.9, II.1.11, II.3.5, II.3.5, II.4.2–.3, V.1.14, VII.2.3, VII.2.8, VII.2.9
Minnesota (MN)	I.4.7, II.1.7, II.3.5, IV.2.9, V.1.14, V.2.10–.14
Mississippi (MS)	I.3.11, I.4.4, I.4.5, I.4.6, II.1.2, IV.3.8, VI.2.1, VI.3.4, VII.3.6, VII.3.7
Missouri (MO)	I.1.8, I.3.11, II.1.2, II.3.11, IV.2.10–.11, VI.3.5, VII.3.6
Montana (MT)	I.1.11, I.4.7, II.1.3, II.3.2, III.1.1, III.1.10–.12, III.3.1, III.3.4–.5, IV.1.1, IV.1.5, V.4.3
Nebraska (NE)	II.1.2, II.3.9
Nevada (NV)	I.4.7, II.1.9, II.2.8, III.1.14, IV.2.7
New Hampshire (NH)	I.4.10, III.3.1, IV.2.12
New Jersey (NJ)	I.4.6, III.3.1, V.1.6–.8, VI.1.5–.10
New Mexico (NM)	I.4.7, I.4.8, II.1.3, II.1.13, II.2.3, III.1.18, III.3.1, IV.1.6, IV.2.6, IV.3.8, V.4.3, VI.3.10
New York (NY)	I.3.9, I.4.6, III.2.4–.5, III.3.1, III.3.5, III.3.13, IV.3.13, V.1.1–.3, V.1.5, V.2.7, V.4.1–.2, VI.1.5–.10, VI.2.6–.7, VI.3.10, VI.4.1–.4, VII.2.2, VII.3.1–.4
North Carolina (NC)	I.1.6.–7, I.3.11, I.4.3, I.4.7, II.1.2, III.3.1, III.3.6, IV.2.8, IV.3.8, V.1.6–.8, VI.2.3, VI.3.5
North Dakota (ND)	I.1.11, I.4.6, II.1.8, III.3.5, IV.1.5, IV.2.9
Ohio (OH)	I.1.11, I.1.16, I.4.3, I.2.10, II.1.7, III.1.15, IV.2.2–.5, VI.1.2, VI.2.4, VI.2.6
Oklahoma (OK)	II.1.2, IV.3.8, VII.3.6
Oregon (OR)	I.1.11, II.1.9, II.3.4, III.3.1, III.3.7, III.3.13, IV.3.9, V.1.14, VII.2.3
Pennsylvania (PA)	I.1.4, II.1.9, III.3.1, III.3.7, IV.2.2–.5, IV.3.8, IV.3.15, V.4.2, VI.1.5–.10, VI.3.9, VI.4.3, V.4.4
Rhode Island (RI)	I.4.10, V.1.9–.11
South Carolina (SC)	I.4.3, IV.2.6
South Dakota (SD)	I.1.11, IV.2.14
Tennessee (TN)	I.3.11, I.4.4, II.1.2, III.3.1, IV.3.8, VII.2.6, VII.3.6
Texas (TX)	I.1.12–.13, I.3.11, I.4.4, II.1.8, II.2.3, II.3.3, III.3.7, IV.1.10, IV.3.6–.8, V.2.10, VI.2.4–.5, VI.2.10, VII.2.10, VII.3.4, VII.3.6
Utah (UT)	I.1.11, I.4.3, II.1.3, II.1.9, III.1.1–.5, III.1.18–.19, III.2.6, III.3.3–.4, III.3.7, IV.3.2, V.4.6

US State	Related Problems
Vermont (VT)	III.1.6, III.3.1, III.3.6, VI.3.1
Virginia (VA)	I.1.6–.8, I.1.5, I.2.10–.12, I.4.5, II.1.2, II.1.8, III.3.1, IV.2.8
Washington (WA)	II.1.8, II.2.9, III.2.7, III.3.1, III.3.8, IV.3.16–.17, VII.2.2–.3
Washington (DC)	I.1.2, I.1.11, II.1.1, II.1.13, III.2.10, IV.2.2, IV.3.9–.11, V.2.6, VI.1.3
West Virginia (WV)	I.1.5, I.1.11, II.1.13, III.3.1, III.3.5, VI.3.1
Wisconsin (WI)	V.1.13–.14, II.1.7, II.3.5, II.4.2–.3, III.1.6, III.3.12, V.1.10, VI.3.10
Wyoming (WY)	I.1.14–.15, II.1.3, II.1.9, II.1.10, II.2.7, II.3.2, II.3.9, III.1.1, III.1.10–.12, III.3.1, III.3.4–.5, IV.1.5, V.4.3

Index

Note. For convenience, Appendix II links mathematical terms to problems. The US states are linked to the problems in Appendix III.

Absecon Lighthouse 128, 134
Acadia National Park 99, 103
Adams J. 13
Adams J.Q. 13
Addams J. 169
Air Force One 16
Alaska Territory 6
Alaska's Bermuda Triangle 62
American Interstate Highway
    System 117
American Samoa 6
Angelou, M. 171
Animal Kingdom 200
animal-shaped buildings 128
Apollo missions 144
Appalachian National Scenic Trail 86
Arches National Park 72

Barnegat Lighthouse 128
Bermuda Triangle 60
Betsy Ross Flag 23
Bill of Rights 30
body mass index (BMI) 183
body surface area (BSA) 183
borough 37
Boston 32
Bridgewater Triangle 63
Brooks Brothers 192
Buck, P.S. 169
buildings 130–132, 136–137
Bunker Hill Monument 157
Bush G.H.W. 13
Bush G.W. 13

Cape Hatteras Lighthouse 128
Cape Lookout Lighthouse 128
Cape Perpetua Scenic Area 54
Carter G. Woodson Home National
    Historic Site 83

Carter, J. 19
Cassatt, M.S. 172
Central Artery/Tunnel 136
Cherries 187
Chesapeake National Historical Park 83
Chimney Tower 47
Chrysler Building 128, 137
Civil War 2
Cleveland, S.G. 13, 16
coldest temperature 48
Colombia River 56
Columbus, C. 1
Concord 32
Constitution of the United States of
    America 30
County-equivalent 37
Cramer's rule 11

Davis, C. 176
Death Valley National Park 48
Denali National Park 45
Devils Tower 47
Dillard's 193
Dinosaur Diamond Prehistoric
    Highway 116
Disney's Animal Kingdom 198

elevation: highest 37, 38, 45, 46
    lowest 37, 38, 49
Elk's Peak 47
Emmy Award 176
Empire State Building 128, 137
EPCOT 200

federal territories 1
Fibonacci number 3
First Amendment 30
First Transcontinental Railway 143
floating bridges 119

Fort McHenry 23

Garfield, J.A. 22
Gauss-Jordan elimination 11
Glacier National Park 97, 98
Glen Canyon National Recreation
    Area 73
Google 150
Goosenecks State Park 82
Grand Canyon National Park 73, 98, 149
Grand Prismatic Spring 53
Grant, U.S. 22
Grapes 187
Grant Circle 73
Great Basin National Park 71
Great Circle Earthworks 721
Great Lakes 55
Guadalupe Mountains National Park
    46, 100
Guam 6
Guinness World Records 151, 189

hamburgers 189
Hamilton Pool 54
Hammerstein II, O. 176
Harding, W.G. 22
Harford 32
Harrison, B. 15, 22
Harrison, W.H. 15, 22
Hayes, R.B. 22
highest elevation 37, 38, 45, 46
highways 116–120;
    widest 117
Hollywood Studios 200
hottest temperature 48

Independence National Historical
    Park 149
Interstate Highway 119
Interstate Highway System 136

jackfruit 188
Jefferson, T. 22

Katy Freeway 117
Kingsley Lake 53

Lake Michigan 61
Lake Pontchartrain Causeway 138

Lake Superior 55
largest hamburger 189
League of Nations 19
Liberty Bell 155
lighthouses 128–129
Lincoln Memorial 83
longest Interstate Highway 119
lowest elevation 37, 38, 49

Martin Van Buren 16
McKinley, W. 22
median 141, 213
Mesa Verde National Park 73, 98
Metropolitan Opera 127
Michigan Triangle 61
Midway Geyser Basin 53
Million Dollar Highway 117
Mississippi River 56
Missouri Territory 3
Mitchell, M. 171
Monroe, J. 22
Monument Valley 73
Morrison, T. 169
Mount Elbert 48
Mount McKinley 45
Mount Rainier National Park 48
Mount Whitney 48
musical Oklahoma! 176

National Anthem of the United States 24
National Millennium trail 87
Nautical mile 98
New York City 55
Niagara Falls State Park 81
Nobel Prize 14, 19, 151, 169, 173
Northwest Territory 38

O'Henry, H. 163
Obama, B. 19
O'Keeffe, G.T. 172
Old Glory flag 23
Olympic Games 150, 153
Oscar Award 176

Pacific Crest Trail 86
Palm Springs Aerial Tramway 137
parish 37
Peaches 188
Pensacola 32

Pentagram 23
Petrified Forest National Park 99
Philippines 6
pizza 189
Platte River 56
Poe, E.A. 163
President of the United States 13
Providence 32
Puerto Rico 6
pumpkins 188

Race Across America 91
Red Rock Amphitheatre 128
Red Rock Scenic Byway 116
Red, White and Blue flag 23
Rocky Mountain National Park 71, 72
Rodgers, R. 176
Roosevelt, F.D. 13, 16

Saguaro National Park 99
Saint Lawrence River 56
San Jacinto Monument 137
Santa Fe 31
sea level 46
Sears 193
Second Continental Congress 24
Sequoia National Park 48, 69
Sierra Nevada 48
Silver Dollar Lake 53
Sondheim, S. 176
Spaceship Earth 198
Stars and Stripes 23, 156
Star-Spangled Banner Flag 23
Star-Spangled Banner National Historic
    Trail 83
state bird 30
state flag 25
state flower 31
state tree 31
Stein Mart 194
strawberries 187
summit 46

Taft, W.H. 22
Tallahassee 32
Tanana River 56
Taylor, T. 22

temperature 48
Tenth Amendment to the US
    Constitution 29
Territory of Hawaii 6
Territory of Wyoming 12
Theodor Roosevelt National Park 98
time zones 30
To Anacreon in Heaven 24
Tony Award 176
topological elevation 46
topological prominence 46
Tree of Life 198
Twain, M. 162
Twelfth Amendment 13
Twenty-Second Amendment 18
Tyler, J. 22

UNESCO World Heritage Sites 149
UNESCO 148
United Nations 19, 148
US department stores 192

Van Buren, M. 16
Vespucci, A. 1
Virgin Islands 6

Waist-to-height ratio 183
Walker, A. 171
Walmart 194
Walt Disney World Resort 197
Washington Crossing the Delaware 156
Washington Monument 137
Washington, G. 13, 22
watermelons 188
Weaubleau crater 56
Western Hemisphere America 1
widest highway 117
Wikipedia 150
Wilson, W. 19, 22
World War II Memorial 155
Wrangell-St. Elias National Park 70

Yellowstone National Park 53, 69
Yosemite National Park 69, 71
YouTube 150
Yukon River 56
Yukon Territory 56

Printed in the United States
by Baker & Taylor Publisher Services